T. Ackermann

E. W. Schmid
G. Spitz
W. Lösch
**Theoretical Physics
on the Personal Computer**

Erich W. Schmid Gerhard Spitz
Wolfgang Lösch

Theoretical Physics
on the
Personal Computer

Second Edition

With 154 Figures

Springer-Verlag Berlin Heidelberg New York
London Paris Tokyo Hong Kong

Professor Dr. Erich W. Schmid
Institut für Theoretische Physik, Universität Tübingen, Auf der Morgenstelle 14,
D-7400 Tübingen, Fed. Rep. of Germany

Dr. Gerhard Spitz
Siemens AG, D-8000 München 70, Fed. Rep. of Germany

Wolfgang Lösch
FB Physik, Universität Essen, Universitätsstr. 5, D-4300 Essen, Fed. Rep. of Germany

Translator: A. H. Armstrong
"Everglades", Brimpton Common, Reading, RG7 4 RY Berks, UK

Title of the original German edition: *Theoretische Physik mit dem Personal Computer*
ISBN 3-540-18310-8 Springer-Verlag Berlin Heidelberg New York
ISBN 0-387-18310-8 Springer-Verlag New York Berlin Heidelberg
© Springer-Verlag, Berlin, Heidelberg 1987

ISBN 3-540-52243-3 2. Auflage Springer-Verlag Berlin Heidelberg New York
ISBN 0-387-52243-3 2nd edition Springer-Verlag New York Berlin Heidelberg

ISBN 3-540-18908-4 1. Auflage Springer-Verlag Berlin Heidelberg New York
ISBN 0-387-18908-4 1st edition Springer-Verlag New York Berlin Heidelberg

Library of Congress Cataloging-in-Publication Data. Schmid, Erich W., 1931– [Theoretische Physik mit dem Personal-Computer. English] Theoretical physics on the personal computer / Erich W. Schmid, Gerhard Spitz, Wolfgang Lösch ; [translator, A. H. Armstrong]. – 2nd ed. p. cm. Translation of: Theoretische Physik mit dem Personal-Computer. Includes bibliographical references (p. ISBN 0-387-52243-3 (U.S. : alk. paper) 1. Mathematical physics–Data processing. 2. Microcomputers. I. Spitz, Gerhard, 1955– . II. Lösch, Wolfgang, 1956– . III. Title. QC20.7.E4S3513 1990 530.1'0285'526–dc20 90-9441

This work is subject to copyright. All rights are reserved, whether the whole or part of the material is concerned, specifically the rights of translation, reprinting, reuse of illustrations, recitation, broadcasting, reproduction on microfilms or in other ways, and storage in data banks. Duplication of this publication or parts thereof is only permitted under the provisions of the German Copyright Law of September 9, 1965, in its version of June 24, 1985, and a copyright fee must always be paid. Violations fall under the prosecution act of the German Copyright Law.

© Springer-Verlag Berlin Heidelberg 1988 and 1990
Printed in Germany

The use of registered names, trademarks, etc. in this publication does not imply, even in the absence of a specific statement, that such names are exempt from the relevant protective laws and regulations and therefore free for general use.

Please note: Before using the programs contained in this book, please consult for technical advice the manuals provided by the respective manufacturer of the computer – and any additional plug-in-boards – to be used. The authors and the publisher accept no legal responsibility for any damage caused by improper use of instructions and programs contained within. The programs appearing here have been tested carefully. Nevertheless we can offer no guarantee for the correct functioning of the programs.

Printing: Weihert-Druck GmbH, D-6100 Darmstadt
Binding: J. Schäffer GmbH & Co. KG., 6718 Grünstadt
2156/3150-543210 – Printed on acid-free paper

Preface to the Second English Edition

The second English edition has given us the opportunity to implement some suggestions made by our readers.

A version of our plot package adapted to the Microsoft FORTRAN-77 compiler (on a diskette supplied by H. U. Zimmermann) had to be mailed to many readers who have a computer without a coprocessor and who therefore have to use a compiler which does not require one. For this reason we have modified the diskette accompanying our book by offering the option to work with Microsoft FORTRAN-77 (version 4.0 and up). In a last-minute effort we were also able to offer the option to use the new FORTRAN/2 compiler by IBM; we thank Dr. von Dehn for his quick help.

The list of personal computers on which we have installed and tested the programs given in the book has become longer. It now includes the IBM PS/2 series, Siemens PC-D and PC-D2, and many other IBM-compatible computers.

Chapter 1 of the book has been rewritten because of the new options. In other chapters we have corrected errors; some paragraphs have been modified to improve understanding.

Tübingen, January 1990 *E. W. Schmid, G. Spitz*

Preface to the First English Edition

We would like to thank Mr. A.H. Armstrong, who translated this book, for his many valuable suggestions and corrections. We also acknowledge a stimulating response from our readers. Mr. J. Peeck sent us a diskette containing the programs modified to run on an ATARI computer. Mr. H.U. Zimmermann sent us diskettes, on which the graphics software of the book is adapted to the requirements of the FORTRAN-77 compiler by MICROSOFT. Readers interested in these adaptations should contact the authors.

Tübingen, January 1988 *E.W. Schmid, G. Spitz*

Preface to the German Edition

This book is based on the lecture course "Computer applications in Theoretical Physics", which has been offered at the University of Tübingen since 1979. This course had as its original aim the preparation of students for a numerical diploma course in theoretical physics. It soon became clear, however, that the course provides a valuable supplement to the fundamental lectures in theoretical physics. Whereas teaching in this field had previously been principally characterised by the derivation of equations, it is now possible to give deeper understanding by means of application examples. A graphical presentation of numerical results proves to be important in emphasizing the physics. Interaction with the machine is also valuable. At the end of each calculation the computer should ask the question: "Repeat the calculation with new data (yes/no)?". The student can then answer "yes" and input the new data, e.g. new starting values for position and velocity in solving an equation of motion.

The programming of a user-friendly dialogue is not really difficult, but time consuming. At the beginning of the course the student therefore constructs only the numerical parts of the programs. The numerical parts are therefore deleted from the programs under consideration, and newly programmed by the student. Later on, the programming of the graphical output and of the dialogue is taught.

In the initial phase of the course several assistants contributed in the preparation of the teaching schedule. We are particularly grateful to Dr. K. Hahn, Dr. R. Kircher, Dr. H. Leeb, Dr. M. Orlowski and Dr. H. Seichter. For suggesting and discussing the example of "the electrostatic lens", we thank Prof. Dr. F. Lenz and Prof. Dr. E. Kasper.

Until 1985 this course had been held at the terminals of the Computer Centre of the University of Tübingen. After the arrival of powerful personal computers the students themselves showed a desire to have the course adapted to these machines. With the welcome cooperation of IBM the FORTRAN programs for all the chapters were rewritten for the personal computer as a diploma task (W. Lösch). In order to make the programs accessible to a wider circle of users this book was produced. We wish you great fun on the personal computer!

Tübingen, October 1987 *E.W. Schmid, G. Spitz*

Contents

1. **Introduction** 1
   1.1 Programming of the Numerical Portions of the Programs 3
   1.2 Programming of the Input and Output 6

2. **Numerical Differentiation and Introduction into Screen Dialogue** 12
   2.1 Formulation of the Problem 12
   2.2 Mathematical Methods 13
   2.3 Programming 14
   2.4 Exercises 19
   2.5 Solutions to the Exercises 19

3. **Numerical Integration** 22
   3.1 Formulation of the Problem 22
   3.2 Numerical Methods 23
       3.2.1 The Trapezoidal Rule 23
       3.2.2 The Simpson Rule 24
       3.2.3 Newton-Cotes Integration 25
       3.2.4 The Gauss-Legendre Integration 25
   3.3 Programming 29
   3.4 Exercises 33
   3.5 Solutions to the Exercises 33

4. **Harmonic Oscillations with Sliding and Static Friction, Graphical Output of Curves** 35
   4.1 Formulation of the Problem 35
   4.2 Numerical Treatment 36
       4.2.1 Transformation of the Differential Equation 36
       4.2.2 The Euler Method 37
   4.3 Programming 37
   4.4 Exercises 41
   4.5 Solutions to the Exercises 41

5. **Anharmonic Free and Forced Oscillations** 43
   5.1 Formulation of the Problem 43
   5.2 Numerical Treatment 44

		5.2.1 Improvement of the Euler Method	44
		5.2.2 The Runge-Kutta Method	46
	5.3	Programming	47
	5.4	Exercises	49
	5.5	Solutions to the Exercises	50
6.	**Coupled Harmonic Oscillations**		53
	6.1	Formulation of the Problem	53
	6.2	Numerical Method	54
	6.3	Programming	55
	6.4	Exercises	57
	6.5	Solutions to the Exercises	57
7.	**The Flight Path of a Space Craft as a Solution of the Hamilton Equations**		59
	7.1	Formulation of the Problem	59
	7.2	Mathematical Methods	63
		7.2.1 Mesh Width Adaptation in the Runge-Kutta Method ..	63
		7.2.2 Coordinate Transformation	66
	7.3	Programming	67
		7.3.1 Hamilton's Equations of Motion	67
		7.3.2 Automatic Mesh Width Adjustment in the Runge-Kutta Method	69
		7.3.3 Coordinate Transformation	71
		7.3.4 Main Program	73
	7.4	Exercises	79
	7.5	Solutions to the Exercises	79
8.	**The Celestial Mechanics Three-body Problem**		81
	8.1	Formulation of the Problem	81
	8.2	Mathematical Method	85
	8.3	Programming	85
	8.4	Exercises	89
	8.5	Solutions to the Exercises	89
9.	**Computation of Electric Fields by the Method of Successive Over-relaxation**		90
	9.1	Formulation of the Problem	90
	9.2	Numerical Method...	92
		9.2.1 Discretisation of Laplace's Equation	92
		9.2.2 The Method of Successive Over-relaxation	93
	9.3	Programming	95
	9.4	Exercises	100
	9.5	Solutions to the Exercises	100

10. The Van der Waals Equation ... 102
10.1 Formulation of the Problem ... 102
10.2 Numerical Method ... 104
10.3 Programming ... 106
10.4 Exercises ... 112
10.5 Solutions to the Exercises ... 113

11. Solution of the Fourier Heat Conduction Equation and the "Geo-Power Station" ... 115
11.1 Formulation of the Problem ... 115
11.2 Method of Solution ... 117
11.3 Programming ... 119
11.4 Exercises ... 121
11.5 Solutions to the Exercises ... 122

12. Group and Phase Velocity in the Example of Water Waves ... 125
12.1 Formulation of the Problem ... 125
12.2 Numerical Method ... 129
12.3 Programming ... 131
12.4 Exercises ... 134
12.5 Solutions to the Exercises ... 134

13. Solution of the Radial Schrödinger Equation by the Fox-Goodwin Method ... 136
13.1 Formulation of the Problem ... 136
13.2 Numerical Method of Solution ... 140
13.3 Programming ... 142
13.4 Exercises ... 144
13.5 Solutions to the Exercises ... 145

14. The Quantum Mechanical Harmonic Oscillator ... 149
14.1 Formulation of the Problem ... 149
14.2 Numerical Method ... 150
14.3 Programming ... 153
14.4 Exercises ... 156
14.5 Solutions to the Exercises ... 156

15. Solution of the Schrödinger Equation in Harmonic Oscillator Representation ... 158
15.1 Formulation of the Problem ... 158
15.2 Numerical Method ... 159
15.3 Programming ... 160
15.4 Exercises ... 163
15.5 Solutions to the Exercises ... 163

16. The Ground State of the Helium Atom by the Hylleraas Method 165
- 16.1 Formulation of the Problem 165
- 16.2 Setting up the State Basis and the Matrix Equation 167
- 16.3 Programming 172
- 16.4 Exercises 180
- 16.5 Solutions to the Exercises 180

17. The Spherical Harmonics 181
- 17.1 Formulation of the Problem 181
- 17.2 Numerical Method 184
- 17.3 Programming 185
- 17.4 Exercises 187
- 17.5 Solutions to the Exercises 188

18. The Spherical Bessel Functions 189
- 18.1 Formulation of the Problem 189
- 18.2 Mathematical Method 191
- 18.3 Programming 192
- 18.4 Exercises 195
- 18.5 Solutions to the Exercises 195

19. Scattering of an Uncharged Particle from a Spherically Symmetric Potential 197
- 19.1 Formulation of the Problem 197
- 19.2 Mathematical Treatment of the Scattering Problem 200
- 19.3 Programming 203
- 19.4 Exercises 206
- 19.5 Solutions to the Exercises 207

References 209

Subject Index 211

1. Introduction

The solution of numerical problems in theoretical physics was until a few years ago the domain of large computers. In recent years there have appeared on the market personal computers which are as powerful as the large computers of the early sixties. Apart from their high computational performance, personal computers offer interactive capabilities and the rapid graphical output of results, which were not available twenty years ago. Personal computers accordingly offer us a wide field of possibilities in education and research.

In education it used to be customary to convey an understanding of the fundamentals of theoretical physics by deduction. Understanding could be consolidated and deepened by the solution of exercises, but the number of analytically soluble examples is unfortunately rather limited. The personal computer now enables us to increase significantly the number of examples by the inclusion of examples which can be solved numerically. The computed results can be graphically displayed in an instructive way. Moreover, in interaction with the computer one can easily vary the physical parameters and the boundary or starting conditions and so become familiar with a whole family of solutions. This clarifies in particular the connection between theory and experiment. The emphasis in teaching is today moving from the derivation of equations towards application.

The personal computer is also a useful tool in theoretical physics research. It performs valuable service in the development and testing of new computer programs. Anybody who has worked for years with large computers, serving many users simultaneously in so-called Time Sharing Mode, knows the joy of having a computer entirely to himself, rather than having to wait in a queue. In applying to complicated scientific studies a program developed on a personal computer, one will still have recourse to large computers. It is most convenient when there is a cable connection between personal computer and large computer.

This book is an attempt to facilitate the introduction of the personal computer into education and research. The programs of the book were developed for computers of the IBM PC/AT type, with mathematical co-processor, 512 KB main memory, hard disc, EGA board and Professional FORTRAN compiler. The programs have now also been tested on computers of the type IBM PC/XT, IBM PS/2, Siemens PC-D and several IBM-compatibles. In most cases no problems arose, provided that the machines were equipped with one of the usual graphics boards. The programs were also tested with two FORTRAN compilers which do not require a mathematical co-processor.

Our decision to use FORTRAN as the programming language may be criticised. There are today several convenient programming languages, and each of these languages has its enthusiastic adherents. The majority of existing numerical programs in physics, however, are written in FORTRAN and this will scarcely change in the next few years. We therefore use FORTRAN, and take the presently most up-to-date form, FORTRAN 77.

In Sect. 1.1 we shall briefly review the methods used for the programming of the numerical portions of the programs. In Sect. 1.2 there follows a description of the libraries which we have developed for the graphical input and output. The reader should skip these two sections at first reading. We recommend, instead, to start with one of the programs of this book and to experiment with it.

First one has to install the software packet of the book. We assume that the reader has at his disposal a personal computer with hard disk and either a Professional FORTRAN compiler or a Microsoft FORTRAN compiler (version 4.0 or later) or an IBM FORTRAN/2 compiler.

Before starting the installation, the compiler and its utility programs must have been installed on the disk. A PATH command must have been issued so that the operating system knows where to look for the compiler. Also, the installation procedure must be informed in which directory the FORTRAN library resides. This is done by entering a SET command. E.g., if the FORTRAN library is in the directory C:\FORT\LIBRARY, one must enter the command

SET LIB = C : \FORT\LIBRARY

before starting the installation procedure.

The first step of the installation is to create a new directory and copy all the files of the book diskette into it. The files named SOURCEG, SOURCEK and SOURCEA contain a graphics library, all training programs of the book and auxiliary data in a compressed form.

Now, one must extract the relevant programs and data from SOURCE and create the graphics library. One does this by entering the command

INSTALL.

The installation procedure now will ask the user for some information. First, one has to select the version of the graphics library and the compiler that are to be used. There are two versions of the graphics library:

— A "standalone version" communicates directly with the graphics hardware, with the help of the built-in software of the computer. It works with most common graphics boards and can be used with all compilers mentioned above. A list of graphics boards that are suitable for this version is given in Sect. 1.2.

– The "VDI version" uses the Graphics Development Toolkit (GDT) and the Virtual Device Interface (VDI) in order to produce graphics. It works only if suitable VDI device drivers are installed on the computer. Since Microsoft does not include an interface to the Graphics Development Toolkit with its compiler, this version can be used only with the Professional FORTRAN or the IBM FORTRAN/2 compiler.

The installation procedure will also ask the user to check once more if he has all the software and hardware needed to run the graphics packet. If one notices that there is still something missing, one can stop the installation procedure. One should read the hints given by the installation procedure *carefully* before deciding to go on.

If the user decides to go on, the installation procedure will extract all necessary programs and data from the file SOURCE. The programs of the graphics library will be compiled and put into the library. This will take some time – several minutes up to half an hour, depending on the speed of the computer. It is not necessary to stay at the computer during this time. When the library has been installed, a test program will be run in order to check if the library works correctly.

After this test, one can start any of the programs of this book, using the command

```
RUN name
```

where "name" is the name of the program without the suffix .FOR.

By the command RUN a program is compiled, linked and executed. The machine code is erased after execution. If the machine code is not to be lost, the command

```
COLI name
```

should be used. Subsequent calling of the program is then achieved simply by input of the program name.

Hints on how to install the library under conditions other than those specified are to be found in Sect. 1.2.

1.1 Programming of the Numerical Portions of the Programs

The FORTRAN 77 language is defined in the ANSI Standard X3.9-1978. We use the language in the form described as the *Full* Standard. Some constructions which are lacking in the Limited Standard or in the older FORTRAN 66 will frequently be used in this book, for example, the possibility of specifying the lower limits of arrays. Only a compiler which uses the full language standard will be able to process the programs of this book without modification.

```
      IS=H/3.D0*(F(A)+F(B))                                110
      DO 20 I=2,N-1,2                                      111
      IS=IS+4.D0/3.D0*H*F(A+(I-1)*H)                       112
20    CONTINUE                                             113
      DO 30 I=3,N-2,2                                      114
      IS=IS+2.D0/3.D0*H*F(A+(I-1)*H)                       115
30    CONTINUE                                             116
```

Fig. 1.1. Coding of equation (1.1) in main program KAP3 (see Fig. 3.3)

```
      IS1=0
      IS2=0
      X=A
      DO 20 I=2,N-1,2
      X=X+H
      IS1=IS1+F(X)
      X=X+H
      IS2=IS2+F(X)
20    CONTINUE
      IS=H/3.D0*(4*IS1+2*IS2+F(A)-F(B))
```

Fig. 1.2. Optimised version of program

In the programming we have taken care that formulae which are printed in the text of the book are directly identifiable in the program. For example, in Chap. 3 the Simpson Rule (3.14)

$$I \approx I_S = \frac{h}{3}f_1 + \frac{4h}{3}f_2 + \frac{2h}{3}f_3 + \frac{4h}{3}f_4 + \ldots + \frac{4h}{3}f_{n-1} + \frac{h}{3}f_n \quad , \qquad (1.1a)$$

with

$$f_i = f(a+(i-1)h) \quad , \quad i=1,2,\ldots,n \quad , \quad h = \frac{b-a}{n-1} \quad , \qquad (1.1b)$$

was transformed into the piece of program shown in Fig. 1.1. In order to economise in computation time one could have taken the factors $2h/3$ and $4h/3$ outside the loop. Also the multiplication in the argument of F could have been avoided, and finally one could have combined the two loops. This would have resulted in something like the piece of program shown in Fig. 1.2. The connection with the original formula (1.1) is not now so immediately recognisable. One has to examine in detail what the piece of program actually does. Moreover, the new variables IS1 and IS2 have to be introduced and declared. Such opportunities for optimising have therefore been taken in this book only when the execution time on the computer is thereby considerably shortened.

With experience in programming, the student will increasingly wish to optimise programs with regard to execution speed and memory requirements. We invite the reader to practise on the programs of this book. However, it should also be said that experienced programmers optimise only at places where a significant saving is to be expected, and emphasize lucidity and readability at all other places.

In the training of programmers today care is taken to educate the student at an early stage in a programming style which is clear and easy to understand. One accordingly breaks the program up into "modules", which carry out clearly defined tasks. Modules in FORTRAN usually consist of subroutines. The module technique not only makes the program clear, it also makes it possible for several programmers to work on the same project with clear allocation of tasks. However, the module technique should not be overdone. Many algorithms, such as the Euler method in Chap. 4, are so simple and short that we have not introduced subroutines for them.

It is often difficult to decide whether one should write a FUNCTION subprogram or a SUBROUTINE. A FUNCTION subprogram is simpler to use in the main program, whereas a SUBROUTINE is more flexible. Oscillator functions or Bessel functions, for example, are usually calculated from recursion formulae, which produce as intermediate results the function values for all the functions with a lower index than the one required. If one uses a SUBROUTINE, then one can transfer all the function values to the main program as an array and possibly use them there. A FUNCTION subprogram, on the other hand, only provides for the transfer of an individual function value. In order to give examples of both possibilities we have chosen the SUBROUTINE form for the oscillator functions in Chap. 14, and the FUNCTION subprogram form for the Bessel functions in Chap. 18.

We have enhanced the clarity of the programs by the indentation of IF-blocks and DO-loops. At its start, an IF-block or a DO-loop is indented by one column; the indentation ceases at the END IF statement and at the concluding CONTINUE statement of the loop, respectively.

Obsolete FORTRAN statements, such as assigned GO TO statements, arithmetic IF or RETURN to labels, were avoided. With ordinary GO TO care was taken that they do not detract too much from the clarity of the program. Thus we have avoided placing the objective of a GO TO statement between another GO TO statement and its objective.

The programs were developed on an IBM PC/AT with the Professional FORTRAN compiler. Later, they were also tested with the Microsoft FORTRAN and IBM FORTRAN/2 compilers.

In one respect of the programming we have deviated from the ANSI Standard for FORTRAN 77: according to the Standard, continuation lines must start in column 6 with a symbol which is found in the FORTRAN list of symbols, hence with a number, a letter or one of a few specified special symbols, such as "+" or "*". All these symbols can lead to confusion if one does not pay attention to the columns when reading a program. In order to avoid such confusion, many programmers start continuation lines with the symbol "&", which is not found in the FORTRAN list of symbols. We have done this also. If a FORTRAN compiler does not accept the symbol "&" in this position, or if "&" is not to be found on the computer keyboard, one will have to use another continuation symbol.

1.2 Programming of the Input and Output

Two kinds of input and output occur in the programs:
1. alphanumeric input/output,
2. graphical input/output.

The alphanumeric input/output uses the instructions and format statements laid down in the ANSI Standard. In the output explicit use was made in many places of the extended ASCII set of symbols (Code Page 437), which is available in the personal computers of IBM and in compatible computers, e.g. for writing framed texts. These texts have to be modified for computers which use a different set of symbols.

In order to present the output more clearly a FORTRAN subroutine called CURSOR was written, which directs the cursor to a specified line and erases the screen below that line. For CURSOR to work correctly, the device driver ANSI.SYS must be loaded, which on personal computers is usually supplied with the DOS diskette. If ANSI.SYS is lacking, CURSOR writes unintelligible symbols on the screen. One can then remove the CURSOR call without damage. CURSOR can be found on the accompanying diskette.

There is still no graphics standard for personal computers which has been so widely accepted that one will find it on the majority of computers. We have therefore written our own graphics library:

1. The "standalone version" communicates directly with the graphics hardware with the help of the built-in software of the computer.
2. The "VDI version" produces graphics by calling subroutines from IBM's Graphics Development Toolkit, which itself uses the Virtual Device Interface (VDI).

The standalone version of the graphics library was developed for the following hardware and software configurations:

— IBM PC, PC/XT, PC/AT, PS/2, or fully compatible machines, or Siemens PC-D
— PC-DOS or MS-DOS from version 2.10
— IBM Professional FORTRAN compiler, Microsoft FORTRAN compiler from version 4.0, or IBM FORTRAN/2 compiler.
— one of the graphics boards listed in Fig. 1.3.

Depending on the compiler, additional hardware may be required. E.g., the Professional FORTRAN compiler, version 1, needs a mathematical co-processor, whereas the microsoft FORTRAN compiler works with or without a co-processor, but requires at least 320 kB RAM. For details, see the user manual of your compiler. The computer should also be equipped with a hard disk. In principle, the graphics library can be used without one, but the installation procedure assumes a hard disk to be present.

Graphics board	Monitor	Graphics Mode	Resolution (points) x y	Colours
Colour Graphics Adapter (CGA)	colour	6	640 200	2
Hercules	monochrome	7	720 348	2
Siemens PC-D	Siemens monochrome	8	640 350	2
Enhanced Adapter (EGA)	standard colour monochrome enhanced colour	14 15 16	640 200 640 350 640 350	16 4 4/16
Multicolour Graphics Adapter (MCGA)	analog colour or gray-scale	17	640 480	2
Video Graphics Array (VGA)	analog colour or gray-scale	18	640 480	16

Fig. 1.3. Graphics configurations and suitable graphics modes

The standalone version is usually able to determine the graphics hardware by itself, and will select an appropriate graphics mode. Figure 1.3 gives a list of allowed hardware combinations and describes the selected graphics modes.

Note:

1. The term "Siemens PC-D" only refers to the original PC-D with Siemens graphics BIOS; the Siemens PC-D2 and PC-D3 are compatible with the IBM PC series, and they are usually equipped with a CGA, EGA or VGA compatible board. The Siemens PC-D is not recognized automatically. When using the Siemens PC-D, the command

    SET HARDWARE = 4

    must be issued before using the graphics packet (see below).
2. The 2 "colours" in graphics mode 6, 7, 8 are dark and bright, the 4 "colours" in graphics mode 15 are dark, bright, blinking and extra bright.
3. An EGA board with enhanced colour monitor will show 4 colours if it has only 64 kByte RAM, otherwise it will show 16 colours.
4. VGA boards that are equipped with an enhanced colour monitor are treated like an EGA board.

5. In graphics modes 14, 16, 17 and 18, the colour palette can be modified by the graphics packet. In the other modes, the colour palette is determined by the built-in software of the computer.

Other graphics boards are allowed only if they can emulate one of the graphics boards listed in Fig. 1.3 in all respects which are relevant to programming. E.g., an IBM Professional Graphics Adapter may be used, but it will be treated like an ordinary Colour Graphics Adapter.

The standalone version of the graphics packet does not check if the video adapter is suitable for graphics; therefore it may never be used with text-only adapters such as the original IBM monochrome graphics adapter.

The graphics packet will not switch between colour and monochrome monitors. If one is using two monitors on a PC, one must switch to the screen suitable for graphics before using the graphics packet.

Before producing graphics on a Hercules adapter, one has to enable graphics output by the command

HGC FULL.

With a CGA or Hercules board, ASCII characters between 128 and 225 (umlauts, block and line graphics characters, Greek letters, mathematical symbols etc.) can only be displayed if they have been loaded into the main memory by the command GRAFTABL. The program GRAFTABL is usually supplied with PC-DOS.

The standalone version of the graphics library uses assembler routines to communicate with the hardware of the computer. The assembler subroutines are supplied in three variants. Each variant is suitable for one of the compilers listed above. Since each compiler uses its own way to pass parameters to subroutines, the assembler subroutines have to be modified if one wants to use a different FORTRAN compiler.

Most FORTRAN subroutines are written in standard FORTRAN 77. The subroutines GOHWO, GOHWI, GOENV, GOLINE, however, communicate with the computer's hardware by calling assembler routines, and they use a number of language extensions specific to the Microsoft FORTRAN, IBM Professional FORTRAN and IBM FORTRAN/2 compilers, such as 2-byte integer and hexadecimal numbers and built-in functions like IOR and IAND. When adapting the graphics packet to a different compiler, these subroutines may have to be modified also.

If the automatic determination of the graphics hardware does not produce the desired result, one should, before running a graphics program, use the command

SET HARDWARE = hwtyp, grmode, xres, yres, ncolrs, cmode, monito

which changes the so-called environment (see DOS manual).

The meanings of the parameters are:

```
hwtyp:    type of hardware in use
          hwtyp = 0 : see below
          hwtyp = 1 : EGA or VGA board
          hwtyp = 2 : CGA board
          hwtyp = 3 : Hercules board
          hwtyp = 4 : PC-D monochrome with Siemens graphics BIOS
          hwtyp = 5 : MCGA board
grmode:   graphics mode (see Fig. 1.3)
xres:     resolution in x-direction (points)
yres:     resolution in y-direction (points)
ncolrs:   number of colours
cmode:    way in which the colours of the colour palette are set
          cmode = 0 : determined by hwtyp
          cmode = 1 : EGA board in 4-colour mode
                      (mode 15 or mode 16 with only 64 kByte of RAM)
          cmode = 2 : EGA or VGA board in 16-colour mode
          cmode = 3 : MCGA board
monito:   type of monitor
          monito = 0 : standard colour or monochrome
          monito = 1 : EGA colour monitor (with EGA or VGA board)
          monito = 2 : MCGA or VGA gray-scale monitor
          monito = 3 : MCGA or VGA colour monitor
```

Usually, it is sufficient to give the hardware type and the desired graphics mode. If one uses an EGA or VGA board, one should specify all parameters, since otherwise the graphics package will assume that the computer is equipped with a standard colour monitor. The other parameters may be useful if one has, e.g., a VGA board with additional special high resolution graphics modes. If one uses hardware type 0, the graphics packet tries to use the BIOS of the computer for producing graphics. This is very slow, but may be useful if one has a nonstandard graphics board, which, however, uses the same BIOS calls as a CGA, EGA, MCGA or VGA adapter in order to produce graphics.

One should never experiment blindly with the **SET HARDWARE** command since some graphics boards may respond unpredictably to random variations of their data. Also, one should not alter the settings of the switches on a graphics board if one does not know exactly what one is doing, even if the switches are accessible without opening the computer. Many monitors can be damaged by incorrect settings of the switches of the graphics board.

In connection with the EGA, VGA, MCGA boards, the standalone version of the graphics packet chooses a colour palette which gives a contrasty image on most monitors. One can change the predefined colours if one enters the command

SET PALETTE = c0, c1, c2, c3, c4, c5, c6, c7, c8, c9, c10, c11, c12, c13, c14, c15

before running the program. Here, c0 is the background colour, c1 the first foreground colour, c2 the second foreground colour, etc.

The colours are coded as follows:

0 = black	4 = red	8 = dark grey	12 = bright red
1 = blue	5 = magenta	9 = light blue	13 = light magenta
2 = green	6 = brown	10 = light green	14 = yellow
3 = cyan	7 = light grey	11 = light cyan	15 = white

By inputting negative numbers between -1 and -63 it is also possible to use the EGA format for the coding of the colours; with this format one can specify the blue, red and green component of each colour more precisely. For more details, one should consult a technical manual on the EGA board.

Since the standalone version of the graphics packet uses interrupts to adjust the graphics board, writes directly to video memory and uses also OUT commands to set the registers of the graphics board, it cannot be easily adapted to run under OS/2 in protected mode. However, it has been tested that it will work in the DOS compatibility box of OS/2.

The VDI version of the graphics packet can work without modification with all graphics boards for which there is a suitable VDI device driver. Such drivers are available for most graphics boards which are used in connection with IBM personal computers and compatibles, and also for a number of other graphics devices such as printers and plotters. For the precise hardware and software requirements, one should consult the user manual of the Graphics Development Toolkit. If several device drivers are available for a graphics board, the one with the highest resolution is usually the most suitable. For the graphics programs given in this book, one should have a resolution of at least 640 points horizontally and 200 or more vertically. The package of the Professional FORTRAN Compiler Version 1.00 includes the Graphics Development Toolkit library.

For the IBM FORTRAN/2 compiler, a GDT library is included with some versions of the IBM Graphics Development Toolkit. We have restricted our tests to these two compilers.

For other compilers, e.g. the Microsoft FORTRAN compiler, it may be possible to buy a Graphics Development Toolkit library as a separate utility. Even if the manufacturer of the compiler does not offer such a library, an independent software company may produce one. With many compilers, a list of utilities available for the compiler is included – a GDT library may be listed there. Such a library may use different procedure names and parameter types than the IBM Graphics Development Toolkit; therefore, subroutine GOHWO may have to be modified. Also GOHWO uses some language extensions which may not be present on all compilers.

The VDI version of the graphics packet can also be used under OS/2 in protected mode if a suitable VDI driver exists. This has been tested for the IBM FORTRAN/2 compiler. The installation procedure on the book diskette, however, is only suitable for MS-DOS or PC-DOS. If one wants to use the VDI version of the graphics packet under OS/2 in protected mode, one has to use the DOS compatibility box in order to install it.

2. Numerical Differentiation and Introduction into Screen Dialogue

2.1 Formulation of the Problem

In this and also in the next chapter we shall not yet be concerned with physical problems. We wish first to learn how to use the computer. At the same time we shall get acquainted with two formulae which we shall be using later. The equations of motion of theoretical physics are as a rule differential equations. In order to be able to solve them numerically, we use numerical differentiation. In what follows we shall obtain a simple approximation for the first derivative, an equally simple approximation for the second derivative and the pertinent correction terms.

Of great importance in all numerical computations is the estimate of the accuracy of the results. A computer calculates with numbers of finite precision. In the use of PROFESSIONAL FORTRAN on the IBM PC/AT we have, for example, the choice between "single precision" and "double precision". With SINGLE PRECISION 4 bytes (= 32 bits) are used for the representation of a (real) floating point number. This gives a precision of about 6 decimal places; the exponent must lie between -37 and $+37$. With DOUBLE PRECISION 8 bytes are used for floating point numbers. This then gives a precision of about 15 decimal places; the exponent must lie between -307 and $+307$. Everything coming after these 6 or 15 decimal places, respectively, is rounded off. It is not surprising that in long computations the rounding errors accumulate. We shall see in this chapter, however, that a rounding error of 100% can arise in a single difference formation. In spite of all the methods for systematic checking of precision, it is useful from the start to accumulate experience and develop an intuitive feeling for situations in which rounding errors become critical. We shall see that with single precision the margin between critical and not critical can be really small, whereas with double precision it is usually large. In the subsequent chapters we shall therefore always compute using double precision.

In the relatively simple examples in this book the mathematical portion of the program will usually be rather short. A larger portion well be taken up with the programming of the screen display. The latter is important; for we do not wish to have to examine the program for the meaning of, for example, the first five numbers which are demanded as input parameters. The computer should ask: "Argument of the function?", "Mesh width H?", and so on. The output of the results also should have explanatory text. This costs a few lines of program, which we shall always need again in similar form; they will be explained in this chapter. To be more precise, we shall first treat the alphanumeric display. The possibilities of graphic presentation well then be explained in later chapters.

2.2 Mathematical Methods

A simple approximation formula for the numerical calculation of the first derivative of a function can be obtained from the definition of the derivative:

$$f'(x) = \lim_{h \to 0} \frac{f(x+h) - f(x)}{h} \ . \tag{2.1}$$

The transition to the limit $h \to 0$ cannot be achieved on the computer. Instead, we can use a small, but finite, value for h. Thus we obtain the formula:

$$f'(x) = \frac{f(x+h) - f(x)}{h} + O(h) \ . \tag{2.2}$$

Unfortunately, this derivation gives us no estimate for the error term $O(h)$. We can, however, obtain an analytic expression for the error term in a simple manner. For this, we expand the function $f(x)$ in a Taylor series; in what follows we shall always assume that the function $f(x)$ is differentiable a sufficient number of times. We obtain:

$$f(x+h) = f(x) + hf'(x) + \frac{h^2}{2}f''(x) + \frac{h^3}{6}f^{(3)}(x) + \cdots \ . \tag{2.3}$$

Substitution of this series in (2.2) gives:

$$O(h) = -\frac{h}{2}f''(x) + \cdots \ . \tag{2.4}$$

The error $O(h)$ is, as already anticipated by the notation, "of order h", which is to say that, for sufficiently small values of h, it is proportional to h.

On the computer one must not choose the mesh width h too small, since below a certain limit the rounding errors become larger than the result. One therefore seeks an approximation formula for which the errors go to zero with a higher power of h. In the present case this can be easily achieved. We subtract from the Taylor series (2.3) for $f(x+h)$ the Taylor series for $f(x-h)$,

$$f(x-h) = f(x) - hf'(x) + \frac{h^2}{2}f''(x) - \frac{h^3}{6}f^{(3)}(x) + \cdots \ , \tag{2.5}$$

and divide the difference by $2h$. As the new approximation formula we obtain:

$$f'(x) = \frac{f(x+h) - f(x-h)}{2h} + O(h^2) \tag{2.6}$$

with the error term

$$O(h^2) = -\frac{h^2}{6}f^{(3)}(x) + \cdots \ . \tag{2.7}$$

As one can see, the error now goes to zero like h^2. Equation (2.6) is called the two-point formula for the first derivative, since neglecting the error term the first derivative is determined from two function values.

The two-point formula for the first derivative suffices for most applications. By comparison of the TAYLOR series for $f(x+h), f(x+2h), f(x-h)$ and $f(x-2h)$ one could also develop a four-point formula, whose error is of order h^4. For the case where the function is known only over a closed interval, special formulae can be obtained for the end-points, e.g. a three-point formula for the lower interval boundary which is of order h^2 and uses the function values at the points $x, x+h$ and $x+2h$. However, these formulae are not suitable for the solution of differential equations, because they lead to instability.

In later chapters we shall also need an approximation formula for the second derivative. Let us look once again at the TAYLOR series for $f(x+h)$ and $f(x-h)$ given in (2.3) and (2.5). One obtains from these the following three-point formula for the second derivative:

$$f''(x) = \frac{f(x+h) + f(x-h) - 2f(x)}{h^2} + O(h^2) \quad , \tag{2.8}$$

with the error term

$$O(h^2) = -\frac{h^2}{12} f^{(4)}(x) + \ldots \quad . \tag{2.9}$$

As with the two-point formula (2.6) for the first derivative, the error here is also of order h^2.

2.3 Programming

We wish to write a program which calculates numerically the first and second derivatives of a function $f(x)$ at a specified point x, and then establishes the errors of the numerical approximations. The numerical approximations are given by (2.6) and (2.8) omitting the error terms. As an example we choose the function $f(x) = \sin(x)$; the exact values for the first and second derivatives are then $\cos(x)$ and $-\sin(x)$. The numerically obtained derivatives will be denoted by $f'(x)$ and $f''(x)$, the exact values by $g_1(x)$ and $g_2(x)$.

The program should be so flexible that the function $\sin(x)$ chosen as an example can be easily replaced by other functions. In FORTRAN this is achieved by use of subroutines of the type FUNCTION for $f(x), g_1(x)$ and $g_2(x)$. In Fig. 2.2 we see these FUNCTION subroutines in the lines 106 to 108, 109 to 111 and 112 to 114.

Since the choice of names in FORTRAN is subject to strict rules we cannot usually adopt the mathematical or physical names of quantities, but must contrive new names for the FORTRAN program. The titles chosen in this chapter are listed in Fig. 2.1. We shall list titles in this way in all chapters of the book.

After this introduction we are now in a position to program formula (2.6). The result is seen in Fig. 2.2 as line 101. In line 102 we find formula (2.8). The sizes of the error terms in (2.6) and (2.8) are obtained by forming the

Mathematical title	FORTRAN title	Mathematical title	FORTRAN title
x	X	$f(x)$	F(X)
h	H	$g_1(x)$	G1(X)
$f'(x)$	Y1	$g_2(x)$	G2(X)
$f''(x)$	Y2	$d_1(x)$	D1
		$d_2(x)$	D2

Fig. 2.1. Titles of the quantities appearing in the main program KAP2

```
        PROGRAM KAP2                                        100
*       (Input: X,H)
        Y1=(F(X+H)-F(X-H))/(2*H)                            101
        Y2=(F(X+H)+F(X-H)-2*F(X))/(H*H)                     102
        D1=ABS(Y1-G1(X))                                    103
        D2=ABS(Y2-G2(X))                                    104
*       (Output: Y1,G1(X),D1,Y2,G2(X),D2)
        END                                                 105

        FUNCTION F(X)                                       106
        F=SIN(X)                                            107
        END                                                 108

        FUNCTION G1(X)                                      109
        G1=COS(X)                                           110
        END                                                 111

        FUNCTION G2(X)                                      112
        G2=-SIN(X)                                          113
        END                                                 114
```

Fig. 2.2. Numerical portion of the main program KAP2 and the FUNCTION subroutines F, G1 and G2

differences
$$d_1(x) = |f'(x) - g_1(x)| \quad \text{and} \quad d_2(x) = |f''(x) - g_2(x)| \ . \tag{2.10}$$

With the titles given in Fig. 2.1 we obtain the lines 103 and 104 of the program in Fig. 2.2.

The mathematical portion of our program is now complete. What is still lacking is the input of the quantities x and h, and also the output of the results. Since FORTRAN is a technical-scientific programming language and was originally conceived for numerical operations, it provides only with some difficulty an optically pleasing layout for input/output on the screen. For example, FORTRAN provides no menus, but allows input/output only line by line.

We wish to demonstrate next how one programs a simple and brief input/output. First of all the program must inform the user what input values it expects; in our case these are the values for x and h. For this we use the WRITE instruction. Then by a subsequent READ instruction the values typed on the input-keyboard are transferred into the program. The two program lines, which we insert in Fig. 2.2 between lines 100 and 101 of the main program KAP2, are:

```
WRITE (*,*) 'X, H = '
READ  (*,*) X, H
```

The values for x and h can be separated from one another either by a comma or a blank.

In order to bring the results onto the screen, we insert between the lines 104 and 105 the following program lines:

```
WRITE (*,*) 'Y1 = ',Y1,' G1(X) ',G1(X),' D1 ', D1
WRITE (*,*) 'Y2 = ',Y2,' G2(X) ',G2(X),' D2 ', D2
```

Accordingly we then obtain the simple program version shown in Fig. 2.3.

```
    PROGRAM KAP2
    WRITE (*,*) 'X, H = '
    READ  (*,*)   X, H
    Y1=(F(X+H)-F(X-H))/(2*H)
    Y2=(F(X+H)+F(X-H)-2*F(X))/(H*H)
    D1=ABS(Y1-G1(X))
    D2=ABS(Y2-G2(X))
    WRITE (*,*) 'Y1 = ', Y1, '   G1(X) = ', G1(X), '   D1 = ', D1
    WRITE (*,*) 'Y2 = ', Y2, '   G2(X) = ', G2(X), '   D2 = ', D2
    END

    FUNCTION F(X)
    F=SIN(X)
    END

    FUNCTION G1(X)
    G1=COS(X)
    END

    FUNCTION G2(X)
    G2=-SIN(X)
    END
```

Fig. 2.3. Complete program with simple input/output

In time, however, we shall no longer be satisfied with this primitive input/output. Particularly for a user who is not completely au fait with the program, it has defects. A good interactive program using the screen as an input/output device should offer, for example, the following facilities:

1. The source program contains a commentary which explains what the program does.
2. At the start of the program a headline or a short text should be displayed.

3. If an input is expected, the program communicates precisely how this should appear. With a recognisably incorrect input the program reacts with a corresponding message and accepts a revised input.
4. Before an impending output of interim or subsidiary results the program asks whether this output is required or should be skipped.
5. The output of results is clear and contains all the essential information for the user.
6. After each output the program asks in what fashion it should proceed.

How can these requirements be realised in the present program?

For 1:   we arrange the commentary portion in the following fashion:

```
*----------------------------------------------------------------------
*     The program KAP2 calculates the first and the second derivative
*     for a function f using the two-point and the three-point method,
*     respectively. The true derivatives are also calculated as
*     functions G1 and G2.
*     --Input quantities--
*     X:      Value x of the argument of the function f, at which the
*             derivatives are to be calculated
*             {REAL variable}
*     H:      Mesh width
*             {REAL variable}
*     --Output quantities--
*     F(X):   Function value of f at the point x
*             {F is a REAL FUNCTION}
*     Y1:     Numerically determined first derivative of f at the
*             point x
*             {REAL variable}
*     Y2:     Numerically determined second derivative of f at the
*             point x
*             {REAL variable}
*     G1(X):  True value of first derivative of f at the point x
*             {G1 is a REAL FUNCTION}
*     G2(X):  True value of second derivative of f at the point x
*             {G2 is a REAL FUNCTION}
*     D1:     Absolute value of the difference between numerically
*             and analytically determined first derivatives
*             {REAL variable}
*     D2:     Absolute value of the difference between numerically
*             and analytically determined second derivatives
*             {REAL variable}
*     --Subprograms called--
*     CURSOR   {SUBROUTINE}
*     F,G1,G2  {REAL FUNCTIONS}
*----------------------------------------------------------------------
```

For 2:   The four lines of program inform the user that the program of Chapter 2 is now standing ready in the computer.

```
WRITE(*,'(T30,A)')              '              '
WRITE(*,'(T30,A)')              ' CHAPTER  2  '
WRITE(*,'(T30,A)')              '              '
WRITE(*,'(T28,A/8(T2/))') 'NUMERICAL DIFFERENTIATION'
```

For 3: Let us consider, for example, how the program might look for the input of an x-value:

```
      WRITE(*,'(T2,A)') 'Please input values for the following quantities:'
1000  CONTINUE
      WRITE(*,'(T2/T2,A)') 'Argument value:  X = '
      READ(*,*,IOSTAT=IOS) X
      IF(IOS.NE.0) THEN
        WRITE(*,'(T2,A/)') 'Input error !'
        GOTO 1000
      ENDIF
```

The IOSTAT parameter takes a non-zero value if the input cannot be unambiguously interpreted by the computer. In that case the request for the input is repeated.

For 4: This point is not applicable for the present program.

For 5: The programming of a pleasing layout for the output requires some effort. We write:

```
      WRITE(*,'(T2/T2,A,F10.5/)')
     &'Function value:    F(X) = ',F(X)
      WRITE(*,'(T2/T2,A/2(T2,A/T2,(3(A,F11.6))/))')
     & '                         |  Numerical  |  Analytic  |  Difference ',
     & '-------------------------                                          ',
     & '1st derivative F'(X)  |  ',Y1,         '  |  ',G1(X),  '  |  ',D1,
     & '-------------------------                                          ',
     & '2nd derivative F"(X)  |  ',Y2,         '  |  ',G2(X),  '  |  ',D2
```

For 6: The question, whether the calculation should be repeated with new input values, is programmed in the following way:

```
3000  CONTINUE
      WRITE(*,'(T2/T2,A)') 'Rerun the program with fresh values for X
     & and H (Y/N) ? '
      READ(*,'(A1)') CONT
      IF(INDEX('Yy ',CONT).NE.0) THEN
        GOTO 1000
      ELSEIF(INDEX('Nn',CONT).EQ.0) THEN
        GOTO 3000
      ENDIF
```

We must declare the variable CONT as a CHARACTER*1 variable at the start of the program. The IF block in conjuction with the INDEX function has the following effect: if we input "Y", "y" or a "blank", or we just press the ENTER key, then there is a jump to the label "1000". If we input something other than "Y", "y", "blank", "N" or "n", the program request us, after a jump to the label "3000", to answer the question again. In other words: if we want to answer the question with "Yes", then we do not have necessarily to input "Y" or "y"; just pressing the ENTER key will suffice. If we wish to answer the question with "No", then we must input "N" or "n".

The complete program, supplemented in this way, is stored on the diskette in the file KAP-02.FOR. We must forgo a more detailed discussion of the input/output facilities of the programming language FORTRAN here, and leave it to the reader to inform himself more fully.

Occasionally we shall use the subroutine CURSOR. Its purpose consists in placing the cursor on the first column of a specified line and erasing all text on the screen from this line downwards. It is called by

CALL CURSOR(LINE)

where the input parameter LINE is an INTEGER expression which can be any chosen number. The subroutine CURSOR is available in the GRAPHIK.LIB library (for the installation of the GRAPHIK.LIB library see Chap. 1).

It remains to mention that the portions of the program which will be used, not for numerical processing, but exclusively for input or output or for man-machine dialogue, are for reasons of clarity set between the two lines

    * >>>>> Input/Output

and

    * <<<<< .

2.4 Exercises

2.4.1 Calculate the derivatives for various x-values and for various mesh widths h. You will find that the accuracy at first increases with decreasing mesh width h, but then decreases again for still smaller mesh widths. In which region must h lie for both derivatives to be accurate to 3 decimal places, and in which region for them to be accurate to 2 decimal places? What is the reason for the decreasing accuracy with very small mesh widths? Which of the two derivatives is the more sensitive? How would the region of h giving a certain error bound vary, if one caused the function $f(x)$ to oscillate more rapidly, e.g. by choosing $f(x) = \sin 100x$?

2.4.2 What happens if one puts $x = 0, \pi/2, \pi, \ldots$? How can these special cases be explained?

2.4.3 The program works in single precision, i.e. floating point numbers are treated as REAL numbers. Modify the program so that it works in double precision. For this you must convert all constants, variables and arrays to DOUBLE PRECISION. Then repeat Exercise 2.1, and find again the region of h giving a certain error bound. Has the region appreciably expanded?

2.4.4 Modifiy the program so that the derivatives are calculated for another function. As simple examples take the logarithmic and exponential functions. Find here again the region of the mesh width h giving a certain error bound.

2.5 Solutions to the Exercises

2.5.1 A typical output is seen in Fig. 2.4, corresponding to $x = 1$ and $h = 0.1$. The region of the mesh width h, for which one obtains 3-figure accuracy for both derivatives, lies between about $h = 0.1$ and $h = 0.01$. If only two-figure accuracy is required, then the region expands and lies between about $h = 0.2$ and $h = 0.003$. The decrease in accuracy with mesh widths chosen too small is a result of the differencing of almost equal numbers. For example,

| Argument value: X = 1. | | | |
| Mesh width: H = 0.1 | | | |

Function value: F(X) = 0.84147

	Numerical	Analytic	Difference
1st derivative F'(X)	0.539402	0.540302	0.000900
2nd derivative F"(X)	-0.840762	-0.841471	0.000709

Fig 2.4. Screen output display for $x = 1$ and $h = 0.1$

if the first four digits of two 6-digit numbers are equal, then their difference can only have an accuracy of two digits (the integers provide an exception, but for these one does not talk of accuracy). The second derivative is more sensitive than the first, since here one is forming the difference of two differences, viz. $[f(x+h) - f(x)] - [f(x) - f(x-h)]$. With a hundredfold more rapid oscillations of $f(x)$, one obtains the same errors if h is chosen a hundred times smaller.

2.5.2 The error for the first derivative vanishes for sufficiently large values of h at $x = \pi/2, 3\pi/2, 5\pi/2, \ldots$. The reason for this is that only derivatives of odd order appear in the error term (2.7). These vanish for the sine function at the specified points. The error also vanishes for the second derivative at $x = 0, \pi, 2\pi, \ldots$, since only the derivatives of even order appear in the error term (2.9). If one chooses h too small, the rounding errors can falsify this result.

2.5.3. On conversion of the program to double precision, the region in which the mesh width h may vary without an appreciable error occurring in the first or second derivative becomes rather large. We obtain 6-digit accuracy for both derivatives when the mesh width lies between $h = 0.001$ and $h = 0.00001$. If one does not know the function $f(x)$ accurately beforehand, i.e. if one does not know how rapidly it varies, then it becomes especially important to have a wide region in which the choice of h is uncritical. Here double precision offers an important advantage with little extra expenditure on programming and computation time.

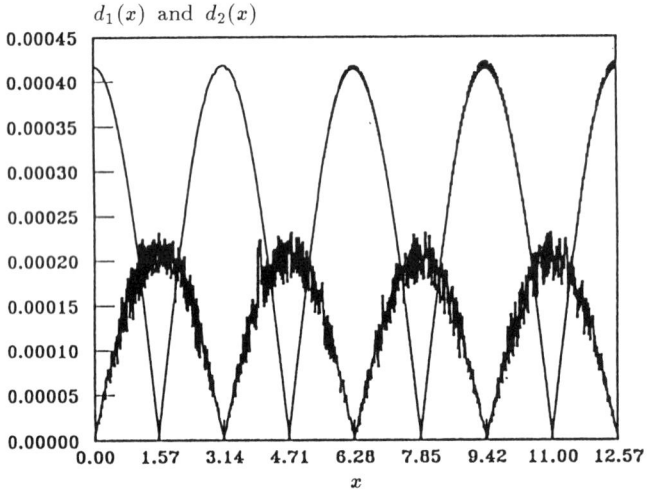

Fig. 2.5. The difference functions $d_1(x)$ and $d_2(x)$ for $h = 0.05$ with single precision calculation

The possibility of graphical output of results will not be discussed until later chapters. Two pictures, however, will be shown here. Figure 2.5 shows the differences $d_1(x)$ and $d_2(x)$, i.e. the errors in the first and second derivatives, plotted as functions of x, for $h = 0.05$. The error of the second derivative especially shows the typical behaviour of statistical noise. Figure 2.6 shows the corresponding curves for double precision calculation. The statistical noise is no longer visible.

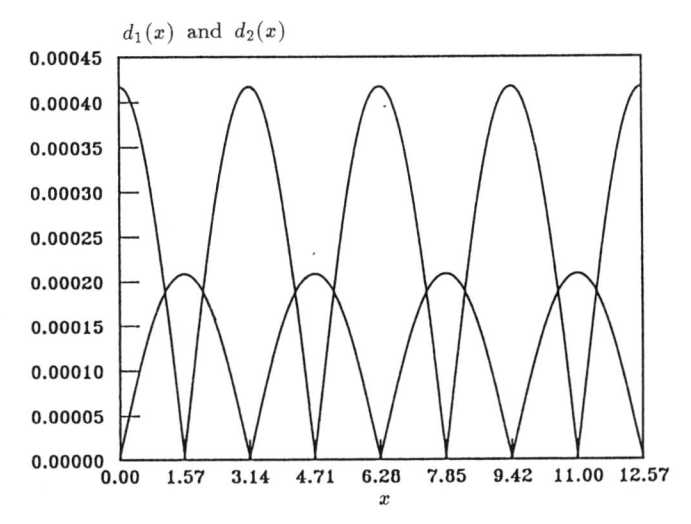

Fig. 2.6. The difference functions $d_1(x)$ and $d_2(x)$ for $h = 0.05$ with double precision calculation

2.5.4 The study of the errors as functions of the mesh width h allows us to set up some rules of thumb. With single precision one will choose h to be, e.g. 1/20 of the interval length in which the function $f(x)$ shows as much variation as $\sin x$ does between 0 and $\pi/2$. With double precision one will choose h instead to be 1/1000 of this interval length, and be confident that a power of ten more or less will make little difference. Here again one sees the advantage of double precision calculations.

3. Numerical Integration

3.1 Formulation of the Problem

The problem of calculating an integral numerically occurs very often in physics. If it involves a one-dimensional integration, and if the integrand is a smooth function, then no difficulties arise with the personal computer. One discretises the integrand over an equidistant mesh and applies a simple rule of integration such as the trapezoidal rule or the Simpson Rule. The required accuracy is achieved by choice of a sufficiently small mesh width.

With multi-dimensional integrations the matter becomes more complicated. Here one has to be very economical with the number of mesh points per dimension and accordingly apply better integration formulae. With the trapezoidal rule the function is linearly interpolated between each pair of neighbouring mesh points, and then integrated. With the Simpson Rule it is quadratically interpolated between each triad of neighbouring points and integrated. The logical development is to combine still more function values by interpolation polynomials of even higher degree and then integrate. This method is known as the Newton-Cotes method. Unfortunately it has drawbacks. The integration weights are very uneven, and with interpolation using more than 7 function values some of the weights become negative.

If one has to economise with the number of mesh points, it pays to replace more extravagant methods such as the Newton-Cotes method by, for example, the method of Gauss-Legendre or methods based on the latter. The trick with the Gauss-Legendre integration is that the integration mesh points are no longer equidistant. The positions of the mesh points are calculated. The additional freedom thereby gained is used to set up an integration rule which gives an exact integration for a polynomial of the highest degree possible. The accuracy achieved by the integration will then depend on whether the integrand can be approximated by a polynomial of high degree. If the integrand is discontinuous, or not differentiable a sufficient number of times, then no gain in accuracy is achieved and the simple trapezoidal rule will give more reliable values for the integral than the more complicated Gauss-Legendre integration.

In this chapter we shall test the trapezoidal rule and the Simpson Rule numerically and compare them with one another. Next we shall carry out a Gauss-Legendre integration and see with how few points one can achieve a comparable accuracy.

3.2 Numerical Methods

3.2.1 The Trapezoidal Rule

We wish to compute the integral of a function $f(x)$ over the interval $[a, b]$. The interval $[a.b]$ is divided into $n - 1$ sub-intervals, so that we obtain n points x_i with the property

$$a = x_1 < x_2 < \ldots < x_n = b \quad . \tag{3.1}$$

We abbreviate the function values at these points as follows:

$$f_i = f(x_i) \quad , \quad i = 1, 2, \ldots, n \quad . \tag{3.2}$$

A simple method for carrying out the numerical calculation of the integral is the so-called trapezoidal rule. One approximates the function between each pair of neighbouring points x_i, x_{i+1} by a straight line. The area of one of the trapezoidal surfaces under the straight lines amounts to

$$F_i = \tfrac{1}{2}(x_{i+1} - x_i)(f_{i+1} + f_i) \quad . \tag{3.3}$$

The sum of the trapezoidal surfaces gives an approximation I_T for the integral

$$\int_a^b f(x)dx = I \approx I_T = F_1 + F_2 + F_3 + \ldots + F_{n-1} \quad . \tag{3.4}$$

In the case of equidistant mesh points the formula (3.4) can be evaluated particularly easily. With the abbreviation

$$h = \frac{b - a}{n - 1} \tag{3.5}$$

it follows that

$$x_i = a + (i - 1)h \quad , \quad i = 1, 2, \ldots, n \quad , \tag{3.6}$$

and one obtains

$$I_T = \frac{h}{2}f_1 + hf_2 + hf_3 + \ldots + hf_{n-1} + \frac{h}{2}f_n \quad . \tag{3.7}$$

An error estimate for the trapezoidal rule is derived in [3.1]. For an equidistant integration mesh of width h the deviation of the approximation I_T from the true value I of the integral is given by

$$|I_T - I| \leq \frac{h^2}{12}(b - a) \max_{a \leq x \leq b}(|f''(x)|) \quad . \tag{3.8}$$

The error is of order h^2, if the second derivative of the integrand exists. If the second derivative does not exist, or if its magnitude is unbounded, other error estimates are available.

3.2.2 The Simpson Rule

Using the trapezoidal rule one can numerically calculate integrals of continuous functions (or functions continuous over parts of the interval). The functions occurring in theoretical physics, however, are usually not only continuous, but also one or more times differentiable. One should expect that advantage can be taken of this property to increase the accuracy of the numerical integration for an equal number of mesh points.

Improved rules of integration can in fact be found. The simplest of them is the Simpson Rule. Instead of joining each pair of points on the curve by a straight line, one constructs a parabolic arc through each triad of points, and integrates over the surfaces under the parabolic arcs.

We wish to restrict ourselves again to equidistant mesh points and consider first the integral over a parabola from -1 to $+1$. We find

$$\int_{-1}^{1} (ax^2 + bx + c)dx = \frac{2a}{3} + 2c \quad . \tag{3.9}$$

The parabola should intersect or touch the integrand $f(x)$ at $x = -1, 0, +1$, i.e. we must have

$$\begin{aligned} f(-1) &= a - b + c \\ f(0) &= c \\ f(1) &= a + b + c \end{aligned} \quad , \tag{3.10}$$

or

$$\frac{2a}{3} + 2c = \frac{f(-1)}{3} + \frac{4}{3}f(0) + \frac{f(1)}{3} \quad . \tag{3.11}$$

We accordingly obtain the approximation

$$\int_{-1}^{1} f(x)dx \approx \frac{2a}{3} + 2c = \frac{f(-1)}{3} + \frac{4}{3}f(0) + \frac{f(1)}{3} \quad . \tag{3.12}$$

We see immediately that this result does not depend on the special choice of integration interval. If one chooses $[a-h, a+h]$ as integration interval, then instead of (3.12) one obtains

$$\int_{a-h}^{a+h} f(x)dx \approx h\left(\frac{f(a-h)}{3} + \frac{4}{3}f(a) + \frac{f(a+h)}{3}\right) \quad . \tag{3.13}$$

By taking a succession of sub-intervals one then obtains the Simpson Rule

$$I \approx I_S = \frac{h}{3}f_1 + \frac{4h}{3}f_2 + \frac{2h}{3}f_3 + \frac{4h}{3}f_4 + \ldots + \frac{4h}{3}f_{n-1} + \frac{h}{3}f_n \quad . \tag{3.14}$$

With the Simpson Rule one always has an even number of interals, i.e. an odd number of points.

With regard to the estimate of error we refer again to [3.1]. If the integrand is four times continuously differentiable, then the deviation of the approximation I_S from the true value I of the integral is given by

$$|I_S - I| \leq \frac{h^4}{180}(b-a) \max_{a \leq x \leq b}(|f^{(4)}(x)|) \quad . \tag{3.15}$$

The error with the Simpson Rule is only of order h^4. As pointed out, the existence of the fourth derivative of the integrand is here assumed. If this does not exist or if it can become very large, then it is safer to apply the trapezoidal rule.

3.2.3 Newton-Cotes Integration

One can generalise the improvement which led from the trapezoidal rule to the Simpson Rule, by interpolating the function between the points by a polynomial of nth degree and integrating it. In this way one obtains the Newton-Cotes method.

The integration weights for equidistant mesh points are tabulated in numerical text-books, e.g. in [3.1]. The integration weights for polynomials up to the 4th degree are reproduced in Fig. 3.1.

Degree	Name	Weight				
1	Trapezoidal rule	h/2	h/2			
2	The Simpson Rule	h/3	4h/3	h/3		
3	3/8 Rule or Pulcherrima	3h/8	9h/8	9h/8	3h/8	
4	Milne Rule	14h/45	64h/45	24h/45	64h/45	14h/45

Fig. 3.1. Integration weights in the Newton-Cotes method

From the 7th degree and beyond a few weights become negative, and the method becomes numerically no longer usable because of increasing rounding error. In practice the higher integration rules are seldom used.

3.2.4 The Gauss-Legendre Integration

The Newton-Cotes integration, especially in the form of the trapezoidal rule or the Simpson Rule, is a very convenient method for the computation of integrals. For functions which can be approximated well by power series in the region of integration, however, Gauss-Legendre integration is significantly more accurate. It operates with fewer points and is therefore better suited for multi-dimensional integrations or when the integrand can be determined only with a high computational effort.

In order to understand the idea behind Gauss-Legendre integration, we shall first consider only the integral of a straight line $g(x)$ over the interval $[-1, 1]$. We know that we can determine this integral exactly using the trapezoidal rule:

$$\int_{-1}^{1} g(x)dx = \int_{-1}^{1} (bx + c)dx = 2c = g(-1) + g(1) \quad . \tag{3.16}$$

Alternatively, however, one can write

$$c = g(0) \quad , \tag{3.17}$$

and hence

$$\int_{-1}^{1} g(x)dx = \int_{-1}^{1} (bx + c)dx = 2c = 2g(0) \quad . \tag{3.18}$$

For exact integration, therefore, a single point suffices, provided we choose this correctly. Gauss-Legendre integration is a generalisation of this idea to polynomials of higher degree. By suitable choice of n mesh points and n integration weights any polynomial up to degree $2n - 1$ is integrated exactly over the interval $[-1, 1]$. We shall give here only a general view and omit the details. A fuller account can be found in numerical text-books, e.g. [3.1]. One first defines an inner product in the space of functions, which are real and analytic in the interval $[-1, 1]$, by

$$\langle f|g \rangle = \int_{-1}^{1} f(x)g(x)dx \quad . \tag{3.19}$$

One can now define a system of orthogonal polynomials with the following properties:

L_n is a polynomial of the nth degree,

$$\langle L_m|L_n \rangle = 0 \quad \text{for} \quad m \neq n \quad . \tag{3.20}$$

By these two conditions the polynomials L_n are uniquely defined to within a normalising factor. The usual normalisation is

$$\langle L_m|L_n \rangle = \delta_{mn} \frac{2}{2n+1} \quad . \tag{3.21}$$

The corresponding polynomials are called Legendre polynomials. The first 3 polynomials are

$$L_0(x) = 1,$$
$$L_1(x) = x,$$
$$L_2(x) = \tfrac{3}{2}x^2 - \tfrac{1}{2} \quad . \tag{3.22}$$

The rest may be found from the Rodrigues formula:

$$L_n(x) = \frac{1}{2^n n!} \frac{d^n}{dx^n}(x^2 - 1)^n. \tag{3.23}$$

The Legendre polynomials and their mathematical properties are discussed in many mathematical works [3.2–4]. One property of the Legendre polynomial which is important in what follows is that it has exactly n zeros in the interior of the interval $[-1, 1]$.

By polynomial division, every polynomial of degree $2n - 1$ or less can be expressed as

$$p_{2n-1}(x) = L_n(x)p_{n-1}(x) + q_{n-1}(x) \quad , \tag{3.24}$$

where $p_{n-1}(x)$ and $q_{n-1}(x)$ are polynomials of degree $n - 1$ or less.
We now have:

$$\int_{-1}^{1} p_{2n-1}(x)dx = \int_{-1}^{1} [L_n(x)p_{n-1}(x) + q_{n-1}(x)]dx$$

$$= \langle L_n | p_{n-1} \rangle + \int_{-1}^{1} q_{n-1}(x)dx \quad . \tag{3.25}$$

The inner product $\langle L_n | p_{n-1} \rangle$ vanishes because L_n is orthogonal to all polynomials of lower degree and p_{n-1} can be expressed as a linear combination of $L_0, L_1, \ldots, L_{n-1}$. It is therefore sufficient to calculate the integral of $q_{n-1}(x)$ in order to obtain the integral of $p_{2n-1}(x)$.

At the zeros $x_1, \ldots, x_n$ of L_n it follows from (3.24) that

$$p_{2n-1}(x_i) = q_{n-1}(x_i) \quad , \quad i = 1, \ldots, n \quad . \tag{3.26}$$

By these n function values q_{n-1} is completely determined, and hence also the integral of q_{n-1}. We wish to compute this integral in a convenient manner. Accordingly let us expand q_{n-1} in Legendre polynomials:

$$q_{n-1}(x) = \sum_{i=0}^{n-1} \alpha_i L_i(x) \quad . \tag{3.27}$$

Because of the orthogonality and normalisation of the Legendre polynomials we have

$$\int_{-1}^{1} q_{n-1}(x)dx = \sum_{i=0}^{n-1} \alpha_i \langle L_0 | L_i \rangle = \alpha_0 \langle L_0 | L_0 \rangle = 2\alpha_0 \quad . \tag{3.28}$$

Instead of an integration problem we now have the problem of determining the coefficient α_0. Since we know the function values of q_{n-1} at the zero points of L_n, we obtain the expansion coefficients appearing in (3.27) from the system of linear equations:

$$\sum_{i=0}^{n-1} L_i(x_k)\alpha_i = q_{n-1}(x_k) \quad , \quad k = 1, \ldots, n \quad , \tag{3.29}$$

which we abbreviate as

$$\sum_{i=0}^{n-1} L_{ki}\alpha_i = q_{n-1}(x_k) \quad , \quad k = 1, \ldots, n \quad . \tag{3.30}$$

Since the Legendre polynomials L_n are linearly independent, one cannot produce any row of the matrix L by a linear combination of the other rows; so L can be inverted. It follows that:

$$\alpha_k = \sum_{i=1}^{n} (L^{-1})_{ki} q_{n-1}(x_i) \quad , \tag{3.31}$$

and hence

$$\int_{-1}^{1} p_{2n-1}(x)dx = 2\alpha_0 = 2 \sum_{i=1}^{n} (L^{-1})_{0i} q_{n-1}(x_i)$$

$$= 2 \sum_{i=1}^{n} (L^{-1})_{0i} p_{2n-1}(x_i) \quad . \tag{3.32}$$

We notice that the matrix L^{-1} is determined by the properties of the Legendre polynomials. It is completely independent of the function to be integrated.

Equation (3.32) already has the form of an integration rule for n mesh points. The points are the zeros of L_n, the integration weights are, apart from a factor 2, the elements of the top row of L^{-1}. One finds the integration weights, together with the zeros of the Legendre polynomials, in published tables, e.g. in [3.4]. The integration weights will be denoted in what follows by w_i. We then obtain the Gauss-Legendre Rule for the integration of a function $f(x)$ over the interval $[-1, 1]$:

$$\int_{-1}^{1} f(x)dx \approx I_G = \sum_{i=1}^{n} w_i f(x_i) \quad . \tag{3.33}$$

The significance of this integration rule lies in the fact that all polynomials up to degree $2n - 1$ are integrated exactly. For analytical functions $f(x)$ one achieves a high degree of accuracy since, in finite intervals, such functions can be approximated well by power series of finite degree. In contrast to the Newton-Cotes rules of high degree, the integration weights in the Gauss-Legendre Rule are neither negative, nor reach extremely high values.

By a linear transformation of the integration variable we can apply the Gauss-Legendre integration to an arbitrary finite interval $[a, b]$:

$$\int_{a}^{b} f(u)du = \frac{b-a}{2} \int_{-1}^{1} f\left(\frac{(b-a)x}{2} + \frac{b+a}{2}\right) dx$$

$$\approx \frac{b-a}{2} \sum_{i=1}^{n} w_i f\left(\frac{(b-a)x_i}{2} + \frac{b+a}{2}\right) = I_G \quad . \tag{3.34}$$

Further development of this method consists in treating also in corresponding manner integrals such as, e.g.

$$\int_{-1}^{1} \frac{f(x)}{\sqrt{1-x^2}} dx \quad \text{(rational Gauss Integration)} \quad ,$$

$$\int_{-\infty}^{\infty} f(x)e^{-x^2} dx \quad \text{(Gauss-Hermite Integration)} \ ,$$

$$\int_{0}^{\infty} f(x)e^{-x} dx \quad \text{(Gauss-Laguerre Integration)} \ .$$

However, we do not wish to go into this here, and refer the reader to textbooks on numerical mathematics, e.g. [3.1].

3.3 Programming

In writing a program for the computation of integrals by the trapezoidal rule, the Simpson Rule and the Gauss-Legendre Rule we shall adhere to the form worked out in Chap. 2. We shall divide the program into two sections; first we shall deal with the trapezoidal rule and the Simpson Rule, and then with Gauss-Legendre integration. For an example the function $f(x) = 1/(2 + x^2)$ will serve. Its integral function is $g(x) = \arctan(x/\sqrt{2})/\sqrt{2}$. The required titles are to be found in Fig. 3.2.

Mathematical title	FORTRAN title	Mathematical title	FORTRAN title
a	A	I	IE
b	B	I_S	IS
h	H	I_T	IT
n	N	I_G	IG
x_i	X(I)	d_S	DS
w_i	W(I)	d_T	DT
$f(x)$	F(X)	d_G	DG
$g(x)$	G(X)		

Fig. 3.2. Notation employed in main program KAP3

As already announced in Chap. 2, we shall compute in double precision. Line 101 in Fig. 3.3 shows us how we can change the usual FORTRAN implicit type declaration for the data type REAL to the data type DOUBLE PRECISION. All constants, variables, fields and functions, whose names begin with one of the letters in the list (A-H, O-Z), assume the data type DOUBLE PRECISION because of the IMPLICIT instruction.

An implicit declaration of type is overruled by an explicit declaration. The variables IE, IS, IG and IT have the data type DOUBLE PRECISION from the instruction in line 102; otherwise they would have had the data type INTEGER, because of their initial letter and the implicit type declaration.

For the mesh points x_i and the integration weights w_i in the Gauss-Legendre integration we use the one-dimensional arrays X and W. We inform

```
      PROGRAM KAP3                                                   100

      IMPLICIT DOUBLE PRECISION (A-H,O-Z)                            101
      DOUBLE PRECISION IE, IG, IS, IT                                102
      PARAMETER (NMAX=24)                                            103
      DIMENSION X(NMAX), W(NMAX)                                     104

*     (Input of A, B and N for the Simpson and trapezoidal rules)

      H=(B-A)/(N-1)                                                  105
      IT=H/2.D0*(F(A)+F(B))                                          106
      DO 10 I=2,N-1                                                  107
      IT=IT+H*F(A+(I-1)*H)                                           108
10    CONTINUE                                                       109
      IS=H/3.D0*(F(A)+F(B))                                          110
      DO 20 I=2,N-1,2                                                111
      IS=IS+4.D0/3.D0*H*F(A+(I-1)*H)                                 112
20    CONTINUE                                                       113
      DO 30 I=3,N-2,2                                                114
      IS=IS+2.D0/3.D0*H*F(A+(I-1)*H)                                 115
30    CONTINUE                                                       116
      IE=G(B)-G(A)                                                   117
      DT=ABS(IT-IE)                                                  118
      DS=ABS(IS-IE)                                                  119

*     (Output of IT,IS,IE,DT and DS)

*     (Input of A,B,N for Gauss-Legendre integration)

      IG=0.D0                                                        120
      DO 40 I=1,N                                                    121
      IG=IG+W(I)*F((B-A)*X(I)/2.D0+(B+A)/2.D0)                       122
40    CONTINUE                                                       123
      IG=(B-A)/2.D0*IG                                               124
      IE=G(B)-G(A)                                                   125
      DG=ABS(IG-IE)                                                  126

*     (Output of IG,IE and DG)

      END                                                            127


      DOUBLE PRECISION FUNCTION F(X)                                 128
      IMPLICIT DOUBLE PRECISION (A-H,O-Z)                            129
      F=1.D0/(2.D0+X*X)                                              130
      END                                                            131

      DOUBLE PRECISION FUNCTION G(X)                                 132
      IMPLICIT DOUBLE PRECISION (A-H,O-Z)                            133
      G=ATAN(X/SQRT(2.D0))/SQRT(2.D0)                                134
      END                                                            135
```

Fig. 3.3. Numerical portion of the main program KAP3 and the FUNCTION subprograms F,G

the program of this by the DIMENSION instruction in line 104. By the PARAMETER instruction in line 103 we assign to the constant NMAX the value NMAX = 24, i.e, the maximum number of mesh points to be used is 24.

The function $f(x)$ and its integral function $g(x)$ are introduced into KAP3 by two FUNCTION subprograms, which now of course have the data type DOUBLE PRECISION (lines 128 to 131 and 132 to 135):

The progamming of the trapezoidal rule (3.7) is very simple. In line 105 the mesh width is calculated from (3.5). In line 106 the contributions at the interval boundaries a and b are determined, and then in the following DO-loop (lines 107 to 109) the contributions at the remaining mesh points are summed. With the Simpson Rule (3.14) we proceed similarly (line 110 to 116). Here we use two DO-loops, one for even and one for odd indices.

The true value of the integral calculated from the integral function $g(x)$

$$I = g(b) - g(a) \tag{3.35}$$

(line 117) serves for the determination of the error arising in the numerical integration (lines 118 and 119),

$$d_T = |I_T - I| \quad \text{and} \quad d_S = |I_S - I| \quad . \tag{3.36}$$

An important part of the program has still not been mentioned, namely the input/output. Since we set out the complete program on the diskette in the datafile KAP-03.FOR, we shall only go into it briefly here. The input of the interval boundaries a and b and the number n of mesh points has to be inserted between lines 104 and 105. In order to simplify this task we shall simply transfer a portion of the input procedure from the program KAP2. We shall do the same in all subsequent chapters. We thus save ourselves work and have the advantage that the input becomes standardised and hence understood at a glance. The output of the integral values I, I_T and I_S and the deviations d_T and d_S follow after line 119 in the form of a table, such as we know already from KAP2.

Now we turn to the second part of the program, which handles the Gauss-Legendre integration. After the initialisation of I_G with the value zero in line 120, the following DO-loop carries out the summation over all the mesh points according to (3.34). In line 124 the sum is finally multiplied by the factor $(b-a)/2$.

As in the first part of the program, we here again compare the integral value I_G determined by the numerical approximation with the value I determined from the integral function (3.35), forming the difference

$$d_G = |I_G - I| \tag{3.37}$$

in lines 125 and 126.

In order to avoid having to input the mesh points x_i and the integration weights w_i by hand, we have stored them in datafiles. For any chosen n such a datafile must exist. Eight such datafiles are to be found on the diskette, for $n = 2, 4, 6, 8, 10, 12, 16$ and 24.

For a more detailed explanation we consider as an example the case $n = 4$. The required data are to be found in the datafile GAUSS-04.DAT and are arranged in the following way:

 0.861136311594053D0   0.347854845137454D0
 0.339981043584856D0   0.652145154862546D0
−0.339981043584856D0   0.652145154862546D0
−0.861136311594053D0   0.347854845137454D0

In each line there are two values: on the left a mesh point and on the right the corresponding integration weight. How do we now tranfer these values into the program? Let us look at Fig. 3.4. After we have input $n = 4$ according to our example (lines 201 to 208), this value 4 is assigned to the internal datafile COUNT in line 209. In FORTRAN an internal datafile is nothing more than a CHARACTER variable. (In contrast to this, for example GAUSS-04.DAT and KAP-03.FOR are

```
9000    CONTINUE                                                         200
        WRITE(*,'(T2/T2,A,I3,A/T2,A)')                                   201
     &  'Number N of points in the interval [A,B] (2 ≤ N ≤',NMAX,')',   202
     &  'Input only values for which datafiles are available! N = '     203
        READ(*,*,IOSTAT=IOS) N                                           204
        IF((N.LT.2).OR.(N.GT.NMAX).OR.(IOS.NE.0)) THEN                   205
          WRITE(*,'(T2,A/)') 'Input error !'                             206
          GOTO 9000                                                      207
        ENDIF                                                            208
        WRITE(COUNT,'(I2.2)') N                                          209
        INQUIRE(FILE='GAUSS-'//COUNT//'.DAT',EXIST=OK)                   210
        IF(.NOT.OK) THEN                                                 211
          WRITE(*,'(T2,A/)')'GAUSS-'//COUNT//'.DAT does not exist !',    212
          GOTO 9000                                                      213
        ENDIF                                                            214
        OPEN(2,FILE='GAUSS-'//COUNT//'.DAT',STATUS='OLD',IOSTAT=IOS)     215
        IF(IOS.NE.0) THEN                                                216
          WRITE(*,'(T2/T2,A)')                                           217
     &    'GAUSS-'//COUNT//'.DAT: datafile opening error !'              218
          STOP                                                           219
        ENDIF                                                            220
        DO 10000 I=1,N                                                   221
          READ(2,*,IOSTAT=IOS) X(I),W(I)                                 222
          IF(IOS.NE.0) THEN                                              223
            WRITE(*,'(T2/T2,A)')                                         224
     &      'GAUSS-'//COUNT//'.DAT: reading error !'                     225
            STOP                                                         226
          ENDIF                                                          227
10000   CONTINUE                                                         228
        CLOSE(2,IOSTAT=IOS)                                              229
        IF(IOS.NE.0) THEN                                                230
          WRITE(*,'(T2/T2,A)')                                           231
     &    'GAUSS-'//COUNT//'.DAT: datafile closing error !'              232
          STOP                                                           233
        ENDIF                                                            234
```

Fig. 3.4. Input of N, X and W for the Gauss-Legendre integration in the main program KAP3

external datafiles.) We have to declare the variable COUNT as a CHARACTER*2 variable at the beginning of the program. By the WRITE instruction in line 209 COUNT gets the value '04'.

In line 210 it is then checked whether the datafile GAUSS-04.DAT exists. Notice here that it is only checked in the current directory. In the case of non-existence the LOGICAL variable OK takes the value .FALSE., and by the following IF-block (lines 211 to 214), after an error message, we are requested to input a new value for n (jumps to Label 9000).

If OK=.TRUE., i.e. the datafile GAUSS-04.DAT is available, then this is opened in line 215, the values are read in by the following DO-loop (lines 221 to 228), and finally in line 229 the datafile is closed again. If an error occurs in opening the datafile, reading it, or closing it, the program stops with the appropriate message.

3.4 Exercises

3.4.1 Compute the integral of $f(x) = 1/(2+x^2)$ for various integration limits and various numbers of mesh points with the trapezoidal rule and with the Simpson Rule. Compare the errors. The errors with the trapezoidal rule and the Simpson Rule are almost the same, if the lower limit of integration is far in the negative and the upper limit far in the positive. Can you give an explanation for this?

3.4.2 Compute the integral of $f(x) = 1/(2+x^2)$ for various integration limits and various numbers of mesh points using Gauss-Legendre integration, and compare the errors with those incurred when using the trapezoidal rule and the Simpson Rule. How much is the error reduced by using Gauss-Legendre integration, for a comparable number of mesh points?

3.4.3 Replace the integrand $1/(2+x^2)$ in your FORTRAN program by other functions, such as, e.g. $\sin x$, $\log x$ etc., and again compare the accuracy obtained with the various integration methods.

3.5 Solutions to the Exercises

3.5.1 The integrand $1/(2+x^2)$ has the form of a bell-shaped curve. If the limits of integration cut off a part of the bell, then the integration with the Simpson Rule is more accurate than that with the trapezoidal rule. If the limits of integration include practically the whole of the bell, then the errors will become about equally large. This can be explained in the following way. If the integrand almost vanishes in the neighbourhood of the two integration limits, then the integration weights at the ends play no part. The other integration weights are all equal to unity for the trapezoidal rule. With the Simpson Rule the result is scarcely altered in the given case if one carries out a second integration with the mesh shifted by h. The mean value of the two results corresponds to an integration with averaged integration weights. The mean value of 4/3 and 2/3 is however equal to unity, i.e. one is integrating with the Simpson Rule and the trapezoidal rule using effectively the same weights.

3.5.2 For a comparable number of mesh points the error with Gauss-Legendre integration is usually a few powers of ten smaller than with integration using the trapezoidal rule or the Simpson Rule. For all three integration rules it is true that one should always consider the integrand carefully before carrying out the numerical integration. For example, if one integrates the function $f(x) = 1/(2+x^2)$ from $x = 0$ to $x = 100$, then with Gauss-Legendre integration using 24 mesh points one obtains only an accuracy of 3 decimal places. On the other hand, if one dissects the integral into a sum of three integrals over the intervals [0,5], [5,30] and [30,100], then one achieves the maximal achievable accuracy of 15 decimal places. The exact choice of partial intervals is not critical. It results simply from the fact that the first interval includes the region in which $f(x)$ is significantly different from zero, whereas in the third interval the integrand almost vanishes. It is rewarding to "play around" and develop a feeling for what the individual integration formulae achieve.

3.5.3 In the FORTRAN program the integrand $f(x)$ and its integral function appear only in the FUNCTION subroutines and in the commentary. Only in these places are changes necessary. With oscillating functions such as, for example, $f(x) = \sin x$, one will find that, even using Gauss-Legendre integration, it is not possible to integrate over many oscillations with few mesh points, without incurring large errors. In this case one must break the integral down into partial integrals.

4. Harmonic Oscillations with Sliding and Static Friction, Graphical Output of Curves

4.1 Formulation of the Problem

In lectures on mechanics simple and coupled harmonic oscillations of point masses are covered in detail. One encounters various types of oscillation, depending on whether there is a driving force or not, and whether frictional forces are taken into account or not. The frictional forces are usually assumed to be forces dependent on the velocity. These first of all play an important role in physics, such as for example in the damping of galvanometers, and secondly they are capable of being handled analytically. Less amenable to analytic calculation are sliding and static friction, since they behave discontinuously at the turning point of an oscillation. Only piecemeal analytic solutions can therefore be obtained. To a computer it presents no difficulty to integrate across the discontinuities. We therefore take as our first physical example the motion of a point mass under the influence of a harmonic restoring force and under the influence of sliding and static frictional forces.

We wish to output the result of the computation graphically, i.e. we wish to see the oscillatory motion as a curve on the screen. We shall become familiar with the necessary dialogue and the relevant programs in this chapter.

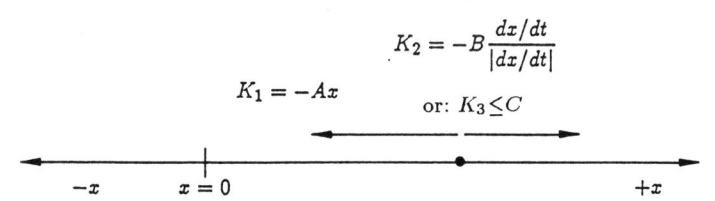

Fig. 4.1. Point mass under the influence of restoring force and frictional force

Figure 4.1 shows the example. A point-like mass $M = 1\,\text{kg}$ moves along the x-axis under the influence of the following forces:

1. An elastic force $K_1 = -Ax$ with $A = 3\,\text{N}\,\text{m}^{-1}$ pushes the point mass towards the origin of the x-axis.
2. A sliding friction force $K_2 = -B(dx/dt)/|dx/dt|$ with $B = 0.5\,\text{N}$ damps the motion ($K_2 = 0$ when $dx/dt = 0$).
3. A static friction force K_3 with $|K_3| \leq C$ and $C = 1\,\text{N}$ holds the point mass still when it is not moving.

The equation of motion is

$$-M\frac{d^2x}{dt^2} - Ax - B\frac{dx/dt}{|dx/dt|} + K_3 = 0 \quad . \tag{4.1}$$

The influence of the static friction force on the oscillatory motion can be indicated at once. When dx/dt becomes equal to zero, then

$$A|x| > C \tag{4.2}$$

must hold, otherwise the point mass remains at rest. With the numerical values for A and C given above, the point mass will remain at rest when

$$dx/dt = 0 \quad \text{and} \quad |x| \le \tfrac{1}{3}\text{m} \quad . \tag{4.3}$$

4.2 Numerical Treatment

4.2.1 Transformation of the Differential Equation

The equation of motion (4.1) has the form

$$\frac{d^2x}{dt^2} = f(x, dx/dt, t) \quad , \tag{4.4}$$

i.e. it amounts to a differential equation of the second order in one dimension.

In order to solve the equation we shall first rewrite it as a system of coupled first order differential equations. For such a system of differential equations there exists a method of solution called the Euler method, which is especially easy to program and which we shall therefore use in this chapter. The conversion is achieved by introducing a second unknown function. The first unknown function is the position x of the point mass as a function of the time t; it will now be denoted by $y_1(t)$. As the second unknown function we introduce the velocity dx/dt of the point mass and denote it by $y_2(t)$.

$$y_1(t) = x(t), \tag{4.5a}$$

$$y_2(t) = \frac{dx(t)}{dt} \quad . \tag{4.5b}$$

We accordingly obtain from (4.4) the system of coupled first order differential equations:

$$\frac{dy_1}{dt} = y_2 \quad , \tag{4.6a}$$

$$\frac{dy_2}{dt} = f(y_1, y_2, t) \quad . \tag{4.6b}$$

The conversion can be generalised to multi-dimensional oscillation equations and to equations for coupled oscillations.

4.2.2 The Euler Method

In order to understand the Euler Method, let us first consider a single differential equation of the type

$$\frac{dy}{dt} = f(y, t) \quad . \tag{4.7}$$

For the derivative on the left-hand side we apply (2.2) and obtain

$$\frac{y(t+h) - y(t)}{h} + O(h) = f(y, t) \quad , \tag{4.8}$$

or

$$y(t+h) = y(t) + hf(y(t), t) + O(h^2) \quad . \tag{4.9}$$

If we ignore the error term $O(h^2)$, then (4.9) becomes a recursion formula which allows us to compute the function $y(t)$, starting from an initial value $y(0)$, in steps of h.

If instead of the differential equation (4.7) we have a coupled system of differential equations of the form

$$\frac{dy_i}{dt} = f_i(y_1, y_2, \ldots, y_n; t) \quad , \quad i = 1, \ldots, n \quad , \tag{4.10}$$

then (4.9) can be generalised, giving

$$y_i(t+h) = y_i(t) + hf_i(y_1(t), \ldots, y_n(t); t) + O(h^2) \quad , \quad i = 1, \ldots, n \tag{4.11}$$

If we ignore the error term, (4.11) becomes a system of recursion formulae, with which the unknown functions $y_1(t)$ can be computed approximately on a mesh of width h. One specifies the initial values $y_1(0), y_2(0), \ldots, y_n(0)$ and computes in the first pass $y_1(h), y_2(h), \ldots, y_n(h)$, in the second pass $y_1(2h), y_2(2h), \ldots, y_n(2h)$, and so on. This recursion is called the Euler Method.

The Euler Method is certainly easy to understand and simple to program, but it is not very accurate and not very stable. At each application of the recursion one moves along the tangents to the solution curves up to the next mesh point. One therefore overshoots the target slightly at each pass. We shall see in our example that the computed oscillation therefore has the tendency to increase its amplitude. If the mesh width h is chosen too large this is clearly visible on the graphical presentation. The use of double precision allows h to be chosen so small that the Euler Method produces good results. The computation time, however, then becomes rather long.

4.3 Programming

First of all the differential equation (4.1), with condition (4.2), has to be put into the form (4.10). Using (4.5) we obtain

$$\frac{dy_1(t)}{dt} = y_2(t), \tag{4.12a}$$

$$\frac{dy_2(t)}{dt} = \begin{cases} 0, & \text{if } y_2(t) = 0 \text{ and } |y_1(t)| \leq \frac{C}{A}, \\ -\frac{A}{M}y_1(t) - \frac{B}{M}\frac{y_2(t)}{|y_2(t)|} & \text{otherwise} . \end{cases} \qquad (4.12b)$$

Comparison with (4.10) yields

$$f_1(y_1(t), y_2(t); t) = y_2(t) , \qquad (4.13a)$$

$$f_2(y_1(t), y_2(t); t) = \begin{cases} 0, & \text{if } y_2(t) = 0 \text{ and } |y_1(t)| \leq \frac{C}{A}, \\ -\frac{A}{M}y_1(t) - \frac{B}{M}\frac{y_2(t)}{|y_2(t)|} & \text{otherwise} . \end{cases} \qquad (4.13b)$$

In order to program a recursion step using the Euler Method, we have to express (4.11) and (4.13) in FORTRAN instructions, which takes only a few lines. We shall accordingly not write a subroutine for the Euler Method, but carry out the whole numerical computation in the main program KAP4, cf. Figs. 4.2 and 4.3.

Physical title	FORTRAN title	Physical title	FORTRAN title
A	A	h	H
B	B	t	T
C	C	t_1	T1
M	M	$y_1(t) = x(t)$	Y1
$y_1(0)$	X0	$y_2(t) = dx/dt$	Y2
$y_2(0)$	V0	$f_1(y_1, y_2; t)$	F1
		$f_2(y_1, y_2; t)$	F2

Fig. 4.2. Notation employed in main program KAP4

Values are assigned to the constants A, B, C and M in a PARAMETER instruction (line 103). The mass M would be an INTEGER quantity because of the implicit type declaration. We must therefore declare it as DOUBLE PRECISION (line 102).

In lines 105 and 106 the two variables Y1 and Y2 are given the initial values X0 and V0. The right-hand sides of the system of differential equations are computed in lines 109 and 110. The Euler recursion (4.11) is carried out in line 112 and 113. In line 111 it is tested whether the motion is brought to a standstill by the static friction. If this occurs, then lines 112 and 113, in which y₁, and y₂ are given new values, are simply bypassed. The condition for the halting of the motion cannot be programmed from (4.13b). Since we are computing with a finite time-step h it is very unlikely that in our integration we shall arrive exactly at the point in time at which the direction of motion reverses and the velocity y_2 vanishes. Accordingly we have to test whether y_2

```
        PROGRAM KAP4                                          100

        IMPLICIT DOUBLE PRECISION (A-H,O-Z)                   101
        DOUBLE PRECISION M                                    102
        PARAMETER (A=3.D0,B=.5D0,C=1.D0,M=1.D0)               103

*       (Input of X0, V0, T1, H)

        T=0.D0                                                104
        Y1=X0                                                 105
        Y2=V0                                                 106

*       (Prepare for graphical presentation, draw axes)

10      CONTINUE                                              107
        IF(T.LT.T1) THEN                                      108
          F1=Y2                                               109
          F2=-A/M*Y1-B/M*SIGN(1.D0,Y2)                        110
          IF(Y2*(Y2+H*F2).LE.0 .AND. ABS(Y1).LE.C/A) GOTO 20  111
          Y1=Y1+H*F1                                          112
          Y2=Y2+H*F2                                          113
20        CONTINUE                                            114
          T=T+H                                               115

*         (Draw arc of curve)

          GOTO 10                                             116
        ENDIF                                                 117

*       (Finish graphical presentation; ask whether to iterate the
        computation)

        END                                                   118
```

Fig. 4.3. Numerical portion of the main program KAP4

has changed sign. By means of a multiplication this is easily programmed. If y_2 changes sign between t and $t+h$, then the product of $y_2(t)$ and $y_2(t+h)$ is negative. If we have arrived exactly at a reversal point, then the product is zero. The second part of the condition (4.13b), namely that $|y_1| \leq C/A$, can be tested directly.

Since we wish to compute y_i not only for one time-step, but for the whole interval from 0 to t_1, we must repeat the lines 109 to 114 until $t \geq t_1$. This is the purpose of the IF-statement in line 108 and the jump in line 116.

We shall not go any further now into the alphanumeric input/output. In this chapter, however, a new form of input/output will be introduced: we shall make use of the graphic capability of the screen. In this and the next chapters we shall get familiar with the graphic subroutines found on the accompanying floppy disk. If the library GRAPHIK.LIB has been correctly installed (see Chap. 1), all the graphic programs are available.

Before one can carry out graphical input/output, the graphical system must be initialised. This is achieved by the call

CALL GOPEN(XMIN,XMAX,XTICK,YMIN,YMAX,YTICK,

XNAME,YNAME,IGRID,ICROSS)

The subroutine GOPEN draws two coordinate axes, subdivides the axes by marks, and labels the marks with numerical values. Also it is possible to label the axes. The individual parameters have the following meaning:

```
*-----------------------------------------------------------------------
*     XMIN:    Smallest value on the x-axis
*              {DOUBLE PRECISION-expression}
*     XMAX:    Largest value on the x-axis
*              {DOUBLE PRECISION-expression}
*     XTICK:   Separation of marks on the x-axis,
*              XTICK= 0.D0: no markings
*              XTICK=-1.D0: automatic determination of markings
*              {DOUBLE PRECISION-expression}
*     YMIN:    Smallest value on the y-axis
*              {DOUBLE PRECISION-expression}
*     YMAX:    Largest value on the y-axis
*              {DOUBLE PRECISION-expression}
*     YTICK:   Separation of marks on the y-axis,
*              YTICK= 0.D0: no markings
*              YTICK=-1.D0: automatic determination of markings
*              {DOUBLE PRECISION-expression}
*     XNAME:   Label on the x-axis
*              {CHARACTER*(*)-expression}
*     YNAME:   Label on the y-axis
*              {CHARACTER*(*)-expression}
*     IGRID:   IGRID=0: Grid lines not visible
*              IGRID=1: Grid lines visible
*              {INTEGER-expression}
*     ICROSS:  ICROSS=0: no coordinate axes
*              ICROSS=1: axes through the origin (X,Y)=(0.,0.)
*              {INTEGER-expression}
*-----------------------------------------------------------------------
```

The graphical input/output is terminated with the call

CALL GCLOSE

All other graphical subroutines must be located between GOPEN and GCLOSE.
The subroutine GLINE draws lines. It is called by

CALL GLINE(X1, Y1, X2, Y2, ICOLOR, ISTYLE)

and draws a straight line from the point (X1,Y1) to the point (X2,Y2). The parameter ICOLOR determines the colour of the line and should have a value between 0 and 15. By ISTYLE one indicates the type of line: $1 =$ continuous, $2 =$ long dashes, $3 =$ dotted, $4 =$ chain dotted (1 point), $5 =$ short dashes and $6 =$ chaindotted (2 points).

Since we construct a curve from a sequence of short straight line segments, we can also use GLINE to draw curves. This method of course has the disadvantage that the most recently drawn point of the curve always has to be specified again by the parameters X1 and Y1. This can be tedious if one has to compute and draw several curves at the same time in a program, such as, for example, the positions and velocities of several moving masses. Accordingly for drawing curves the subroutines GMOVE and GDRAW are available. The subroutine GMOVE fixes the starting point (X0,Y0) for one of the curves to be drawn, and is called by

CALL GMOVE(NUMBER, X0, Y0, ICOLOR, ISTYLE)

The starting points for up to 20 curves can be specified, which are distinguished by the parameter NUMBER. The significance of the parameters ICOLOR and ISTYLE is the same as it is for GLINE. The drawing of the curve is carried out by the subroutine GDRAW. The call

CALL GDRAW(NUMBER, X, Y)

causes a straight line to be drawn to the current point (X,Y). One can apply GDRAW an arbitrary number of times in succession in order to draw a curve in this way.

The routine GCOLOR enables the user to select the colour for the graphical presentation. After the call

CALL GCOLOR(NOTE, ICOLOR)

there appears on the screen a palette of colours, where each colour is labelled with a letter. Depending on the number of colours which can be displayed on the screen, up to 16 colours are offered, where the background colour is always shown with the letter "A". By inputting the corresponding letter the parameter ICOLOR is assigned a value chosen from the list $A \to 0, B \to 1, \ldots, P \to 15$. This value is then returned to the calling program. If the screen can show only one foreground and one background color, then GCOLOR automatically choses the foreground colour. The parameter NOTE is used for outputting, in additon to the colour palette, a one-line commentary which is passed as a CHARACTER*(*)-expression.

The complete program is stored on the diskette in the file KAP-04.FOR.

4.4 Exercises

4.4.1 Examine the stability of the Euler method of solution by varying the integration mesh width h. In which region does one obtain a stable solution? What happens when h is chosen too large?

4.4.2 Compute the displacement of the point mass as a function of time for various initial values of the displacement and the velocity. What effect does the frictional damping have? Where does the point mass come to rest?

4.4.3 Follow an oscillatory motion through several periods, and determine the period of oscillation. Does the sliding friction have an influence on the period of oscillation? How can the result of your investigation be explained?

4.5 Solutions to the Exercises

4.5.1 A satisfactory accuracy is obtained with integration mesh widths which are smaller than about 0.001 s. The program would allow still smaller widths, but the computation time then becomes very long. If the width is greater than $h = 0.03$ s, then the amplitude of oscillation increases in spite of frictional damping.

4.5.2 The oscillation amplitude diminishes from maximum to maximum each time by an equal amount. The point mass comes to rest as a rule not at $x = 0$, but gets stuck with a displacement $|x| \leq 1/3$ m. Figure 4.4 shows the screen output of an oscillation curve with initial displacement 3 m and initial velocity 2 m/s.

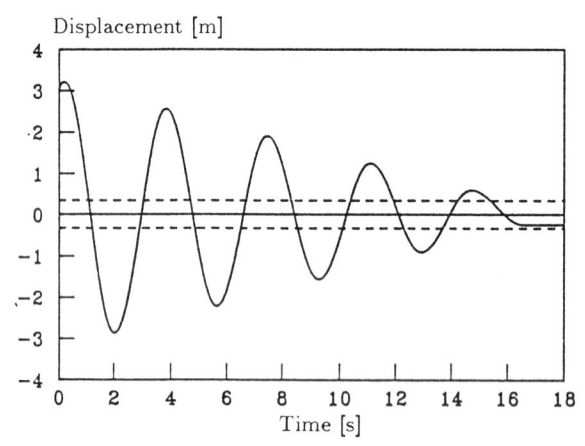

Fig. 4.4. Oscillation curve with initial displacement 3 m and initial velocity 2 m/s

4.5.3 The harmonic oscillation without friction would have a period of $\tau = 2\pi\sqrt{(M/A)} = 3.63$ s. Within the computing and reading accuracy one finds this value also for oscillation with sliding and static friction. This can be understood in the following way. The static friction can affect the period of oscillation only when the point mass comes to rest, since it has no effect at other times. In order to understand the influence of the sliding friction, one must look more closely at the equation of motion(4.1). For positive velocities, and hence for motion in the direction of the positive x-axis, the sliding friction force has the constant value $-B$. In the opposite velocity direction it has the constant value $+B$. One can combine these constant values with the second term of the equation and obtain

$$-Ax - B\frac{dx/dt}{|dx/dt|} = \begin{cases} -A(x + B/A), & \text{when } dx/dt \text{ is positive} \\ -A(x - B/A), & \text{when } dx/dt \text{ is negative} \end{cases} \quad (4.14)$$

The sliding friction force has the effect that the centre of force is displaced from $x = 0$ to $x = -B/A$ or $x = +B/A$, according to which way the point mass is moving. Such a shift of the centre of force, however, has no influence on the period of oscillation. The amplitude merely decreases from reversal point to reversal point by the value $2B/A$.

5. Anharmonic Free and Forced Oscillations

5.1 Formulation of the Problem

An exactly harmonic potential seldom occurs in nature; a small anharmonicity is almost always present. In analytic calculations such perturbation terms present considerable difficulties. In numerical calculations on the computer, on the other hand, it makes scarcely any difference whether the potential is harmonic or anharmonic. In what follows we shall again consider the one-dimensional motion of a point mass with mass $M = 1\text{kg}$. Friction will be ignored. The potential of the restoring force has the form

$$V(x) = A \frac{|x|^{B+1}}{(B+1)} \quad . \tag{5.1}$$

The restoring force is then

$$K(x) = -A|x|^B \frac{x}{|x|} \quad . \tag{5.2}$$

If we also take into account a harmonic driving force the equation of motion becomes

$$-M\frac{d^2x}{dt^2} - A|x|^B \frac{x}{|x|} + C\cos\omega t = 0 \quad . \tag{5.3}$$

Figure 5.1 shows a few potential forms. For $B \to 0$ we get an inclined plane in each of the positive and negative x-directions. When the driving force vanishes this case can be treated quite simply even by analytic means. One would have the solution of the free-falling body both for positive and negative x-values and could match the solutions at the origin. The oscillation period would then increase with the square root of the amplitude. When $B = 1$ we again have the harmonic case, and in the absence of the driving force the oscillation period is independent of the amplitude. For large values of B we approach the case of rigid reflecting walls, and the oscillation period becomes shorter with increasing amplitude.

It is interesting to study the effect of the driving force. It pushes the mass to and fro with a fixed predetermined frequency ω. What will happen if we start in the rest position and let the force operate on the mass? The driving force can operate effectively, i.e. impart energy, only if its frequency corresponds well with the oscillation frequency. For an anharmonic oscillation this will occur

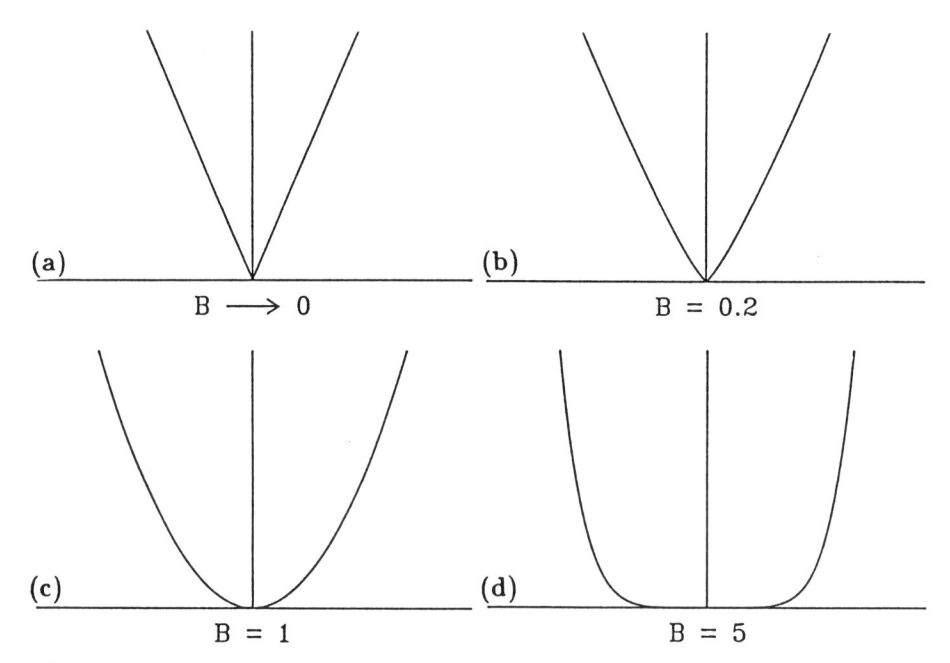

Fig. 5.1a-d. The potential $V(x)$ for values of B

only when a certain amplitude is reached. Let us leave it to the computer to show us what will happen!

5.2 Numerical Treatment

5.2.1 Improvement of the Euler Method

The exercises in Chap. 4 have shown that the Euler method requires long computation times, in order to achieve accurate results. If the motion is complicated, e.g. by strong variations in the curvature of the solution, then the Euler method is no longer useful. We must therefore look for a better method.

The Runge-Kutta method, which we shall use in this and the following chapters, is somewhat more difficult to understand than the Euler method. As a bridge to the Runge-Kutta method, we shall therefore discuss first a possible improvement of the Euler method.

We consider again a first order differential equation of the type

$$\frac{dy}{dt} = f(y,t) \quad . \tag{5.4}$$

In order to be able to apply the technique of the Taylor expansion, we shall assume in what follows that $f(y,t)$ is differentiable a sufficient number of times with respect to y and t. Then, writing down the first few terms of the Taylor expansion of $y(t)$ and applying (5.4), we obtain

$$y(t+h) = y(t) + h\frac{dy(t)}{dt} + \frac{h^2}{2}\frac{d^2y(t)}{dt^2} + O(h^3)$$
$$= y(t) + hf(y(t),t) + \frac{h^2}{2}\frac{df(y(t),t)}{dt} + O(h^3) \quad . \tag{5.5}$$

To differentiate $f(y,t)$ with respect to t we apply (2.2)
$$\frac{df(y(t),t)}{dt} = \frac{f(y(t+h),t+h) - f(y(t),t)}{h} + O(h) \quad , \tag{5.6}$$

and replace the quantity $y(t+h)$ in (5.6) using (4.9). Substitution in (5.5) leads after some simple manipulation to

$$y(t+h) = y(t) + h\frac{f(y(t) + hf(y(t),t), t+h) + f(y(t),t)}{2} + O(h^3) \quad . \tag{5.7}$$

We thus obtain an improved recursion formula, whose error term is now only of order h^3. One sees immediately where the improvement lies, compared with (4.9): instead of the gradient of the solution curve at the mesh point t, we now use the mean value of the gradients at the mesh points t and $t+h$. In order to specify the gradient at the mesh point $t+h$, one actually needs to know already the solution curve at this point. Since this is not the case, it is substituted by the approximation for $y(t+h)$ obtained from the usual Euler formula (4.9). The error arising from this is of order h^3 and is accordingly of the same order as the error already incurred in (5.5).

For practical computation one often introduces the following abbreviations:
$$k^{(1)} = f(y(t),t),$$
$$k^{(2)} = f(y(t) + hk^{(1)}, t+h) \quad . \tag{5.8}$$

In this abbreviated notation the formula for the improved the Euler method is as follows:
$$y(t+h) = y(t) + \frac{h}{2}\left(k^{(1)} + k^{(2)}\right) + O(h^3) \quad . \tag{5.9}$$

The generalisation to several coupled equations follows immediately, since we have nowhere explicitly used the fact that $y(t)$ is a scalar function. We may regard (5.4) as a system of equations for vector functions
$$\frac{dy_i(t)}{dt} = f_i(y_1(t), y_2(t), \ldots, y_n(t); t), \quad i = 1, \ldots, n \quad . \tag{5.10}$$

From (5.8) we then write
$$k_i^{(1)} = f_i(y_1(t), y_2(t), \ldots, y_n(t); t) \quad ,$$
$$k_i^{(2)} = f_i(y_1(t) + hk_1^{(1)}, y_2(t) + hk_2^{(1)}, \ldots, y_n(t) + hk_n^{(1)}; t+h) \quad , \tag{5.11}$$

and hence from (5.9)
$$y_i(t+h) = y_i(t) + \frac{h}{2}\left(k_i^{(1)} + k_i^{(2)}\right) + O(h^3), \quad i = 1, \ldots, n \quad . \tag{5.12}$$

5.2.2 The Runge-Kutta Method

The improved Euler method is better than the ordinary Euler method by one order in h. One can carry the improvement further, raising the order in h by the addition of further terms of the Taylor series (5.5). The best known of the methods obtained in this way is the Runge-Kutta method. Besides the gradients at the beginning and the end of the interval, it also uses the gradient at the middle of the interval, the values of the solution functions $y_i(t)$ at the middle and end of the interval being suitably extrapolated from the values at the beginning of the interval. Both in theoretical derivation and in practical application the Runge-Kutta method is similar to the improved Euler method. The error term in the recursion formula, however, is of order h^5, i.e. the Runge-Kutta method gives an improvement of a further two orders in h compared with the improved Euler method. What this means in practice, we shall see in the exercises.

The Runge-Kutta method uses four gradients, which one denotes by $k^{(1)}$ to $k^{(4)}$. The quantities $k^{(2)}$ and $k^{(3)}$ are gradients at the middle of the interval calculated in different ways. Their average value is used. Using the notation of (5.11), one calculates the gradients $k^{(1)}$ to $k^{(4)}$ as follows:

$$k_i^{(1)} = f_i(y_1(t), \ldots, y_n(t); t) \quad, \tag{5.13a}$$

$$k_i^{(2)} = f_i\left(y_1(t) + \frac{h}{2}k_1^{(1)}, \ldots, y_n(t) + \frac{h}{2}k_n^{(1)}; t + \frac{h}{2}\right) \quad, \tag{5.13b}$$

$$k_i^{(3)} = f_i\left(y_1(t) + \frac{h}{2}k_1^{(2)}, \ldots, y_n(t) + \frac{h}{2}k_n^{(2)}; t + \frac{h}{2}\right) \quad, \tag{5.13c}$$

$$k_i^{(4)} = f_i\left(y_1(t) + hk_1^{(3)}, \ldots, y_n(t) + hk_n^{(3)}; t + h\right) \quad. \tag{5.13d}$$

The recursion formula of the Runge-Kutta method then becomes

$$y_i(t+h) = y_i(t) + h\left(\frac{k_i^{(1)}}{6} + \frac{k_i^{(2)}}{3} + \frac{k_i^{(3)}}{3} + \frac{k_i^{(4)}}{6}\right) + O(h^5) \quad,$$
$$i = 1, \ldots, n \quad. \tag{5.14}$$

For the derivation of the method we refer to [5.1].

The Runge-Kutta method was for a long time the most frequently used and best method for the numerical solution of ordinary differential equations, and even today it is often used. In recent year, however, a number of other methods, especially so-called predictor-corrector methods, have achieved prominence. Since the Runge-Kutta method is easy to understand and simple to program, we shall use it here and in the following chapters on theoretical mechanics.

5.3 Programming

Let us once again consider the system of differential equations (4.12) presented in Chap. 4. By a similar procedure we obtain from (5.3) the system of equations.

$$\frac{dy_1(t)}{dt} = y_2(t) , \qquad (5.15a)$$

$$\frac{dy_2(t)}{dt} = -\frac{A}{M}|y_1(t)|^B \frac{y_1(t)}{|y_1(t)|} + \frac{C}{M} \cos \omega t . \qquad (5.15b)$$

The right-hand sides of the system of differential equations (5.10) are accordingly

$$f_1(y_1(t), y_2(t); t) = y_2(t) , \qquad (5.16a)$$

$$f_2(y_1(t), y_2(t); t) = -\frac{A}{M}|y_1(t)|^B \frac{y_1(t)}{|y_1(t)|} + \frac{C}{M} \cos \omega t . \qquad (5.16b)$$

For the evaluation of the recursion formula (5.14) of the Runge-Kutta method, let us write a subroutine which can be applied to any system of differential equations of the form (5.10). We call this subroutine RUNGE and present it in Fig. 5.3. The notation used is set out in Fig. 5.2.

Mathematical title	FORTRAN title	Mathematical title	FORTRAN title
n	N	$k_i^{(1)}$	K1(I)
h	H	$k_i^{(2)}$	K2(I)
t	T	$k_i^{(3)}$	K3(I)
$y_i(t)$	Y(I)	$k_i^{(4)}$	K4(I)

Fig. 5.2. Notation employed in subroutine RUNGE

As input parameters RUNGE requires the dimension n of the system of equations, the mesh width h, the time t and the initial values $y_i(t)$. The parameter list of RUNGE also contains the subroutine DGL. The program from which RUNGE is called must substitute for the parameter DGL the name of a subroutine declared as EXTERNAL, which calculates the right-hand sides of the system of equations (5.10), $f_i(y_1, y_2, \ldots, y_n; t)$ for the specified values $y_1, y_2, \ldots, y_n, t$.

Since RUNGE requires the DOUBLE PRECISION auxiliary arrays K1 to K4 and YAUX, one first tests whether these quantities are adequately dimensioned to enable the system of equations to be solved, otherwise the program is stopped (line 106).

```
      SUBROUTINE RUNGE(N,H,T,Y,DGL)                              100
      IMPLICIT DOUBLE PRECISION (A-H,O-Z)                        101
      DOUBLE PRECISION K1, K2, K3, K4                            102
      PARAMETER (NMAX=10)                                        103
      DIMENSION Y(NMAX), YAUX (NMAX)                             104
      DIMENSION K1(NMAX), K2(NMAX), K3(NMAX), K4(NMAX)           105
      IF(N.GT.NMAX) STOP 'Error in RUNGE: raise NMAX !'          106
      CALL DGL(T,Y,K1)                                           107
      DO 10 I=1,N                                                108
         YAUX(I)=Y(I)+H*K1(I)/2                                  109
10    CONTINUE                                                   110
      CALL DGL(T+H/2, YAUX ,K2)                                  111
      DO 20 I=1,N                                                112
         YAUX(I)=Y(I)+H*K2(I)/2                                  113
20    CONTINUE                                                   114
      CALL DGL(T+H/2, YAUX ,K3)                                  115
      DO 30 I=1,N                                                116
         YAUX(I)=Y(I)+H*K3(I)                                    117
30    CONTINUE                                                   118
      CALL DGL(T+H   , YAUX ,K4)                                 119
      DO 40 I=1,N                                                120
         Y(I)=Y(I)+H*(K1(I)/6+K2(I)/3+K3(I)/3+K4(I)/6)           121
40    CONTINUE                                                   122
      END                                                        123
```

Fig. 5.3. Subroutine RUNGE

The call of DGL in line 107 causes the $k_i^{(1)}$ to be calculated from (5.13a). Subsequently the auxiliary array YAUX takes the values $y_i(t) + hk_i^{(1)}/2$, which are required for the calculation of the $k_i^{(2)}$ (lines 108–110). In line 111 the $k_i^{(2)}$ are calculated from (5.13b). The calculations of $k_i^{(3)}$ and $k_i^{(4)}$ follow the same way. Finally in the last DO-loop (lines 120 to 122) the $y_i(t+h)$ are calculated from (5.14).

When called from the main program KAP5, RUNGE gets the subroutine DGL5 as an actual parameter. DGL5 has to calculate the functions occurring on the right-hand sides of (5.15). Because of the call in RUNGE, only T, Y and F must stand in the parameter list of DGL5. The additional parameters A, B, C, M and ω appearing in (5.15b) are transferred from the main program in a COMMON block. The equations (5.16a) and (5.16b) are programmed in lines 205 and 206.

Physical title	FORTRAN title	Physical title	FORTRAN title
A	A	t	T
B	B	$y_i(t)$	Y(I)
C	C	$f_i(y_1(t), y_2(t); t)$	F(I)
M	M		
ω	OMEGA		

Fig. 5.4. Notation employed in subroutine DGL5

```
      SUBROUTINE DGL5(T,Y,F)                                            200
      IMPLICIT DOUBLE PRECISION (A-H,O-Z)                               201
      DOUBLE PRECISION M                                                202
      COMMON /DAT/ A, B, C, M, OMEGA                                    203
      DIMENSION Y(2), F(2)                                              204
      F(1)=Y(2)                                                         205
      F(2)=-A/M*(ABS(Y(1)))**B*SIGN(1.D0,Y(1))+C/M*COS(OMEGA*T)         206
      END                                                               207
```

Fig. 5.5. Subroutine DGL5

Physical title	FORTRAN title	Physical title	FORTRAN title
A	A	h	H
B	B	t	T
M	M	t_1	T1
C	C	$y_1(t) = x(t)$	Y(1)
ω	OMEGA	$y_2(t) = dx/dt$	Y(2)
$y_1(0)$	X0	$f_1(y_1, y_2; t)$	F(1)
$y_2(0)$	V0	$f_2(y_1, y_2; t)$	F(2)

Fig. 5.6. Notation employed in main program KAP5

The notation employed in the main program KAP5 is shown in Fig. 5.6. One recognises the similarity to the notation in KAP4. Since the subroutines RUNGE and DGL5 use arrays for storing y_i and f_i, we have now instead of the variables Y1, Y2, F1 and F2 the DOUBLE PRECISION arrays Y and F (line 303 in Fig. 5.7). The variables A, B, C, M and OMEGA are declared in the COMMON block named DAT, since they are required also in subroutine DGL5 (line 305).

The recursion loop of KAP5 (lines 310–319) is constructed in almost the same way as that of KAP4. The IF-block allows us to choose, using the variable METHOD, whether the Euler or the Runge-Kutta method should be employed. If METHOD=1, the Euler method is used. Since we have already introduced the subroutine DGL5 for the calculation of the $f_i(y_1, y_2; t)$ in the Runge-Kutta method, we use it also in the Euler method (line 313). Otherwise the Euler method is programmed as in KAP4. If METHOD=2, the subroutine RUNGE is called.

The complete program is stored on the diskette in the file KAP-05.FOR

5.4 Exercises

5.4.1 Test the accuracy of the Runge-Kutta method in the case of a free harmonic oscillation $B = 1$ and in the case of a strongly anharmonic oscillation ($B = 5$). *Hint*: If the mesh width h is too big, very strong forces can lead to exponent overflow. In this case abort the program and restart it with new parameters!

5.4.2 Compare the solution curves of the anharmonic oscillations calculated with various values for B.

```
        PROGRAM KAP5                                              300

        IMPLICIT DOUBLE PRECISION (A-H,O-Z)                       301
        DOUBLE PRECISION M                                        302
        DIMENSION Y(2),F(2)                                       303
        EXTERNAL DGL5                                             304
        COMMON /DAT/ A, B, C, M, OMEGA                            305

*       (Input of X0, V0, A, B, C, OMEGA, H, T1, METHOD)

        M=1.D0                                                    306
        T=0.D0                                                    307
        Y(1)=X0                                                   308
        Y(2)=V0                                                   309

*       (Initialise drawing, draw system of coordinates)

10      CONTINUE                                                  310
        IF(T.LE.T1) THEN                                          311
         IF(METHOD.EQ.1) THEN                                     312
          CALL DGL5(T,Y,F)                                        313
          Y(1)=Y(1)+H*F(1)                                        314
          Y(2)=Y(2)+H*F(2)                                        315
         ELSE                                                     316
          CALL RUNGE(2,H,T,Y,DGL5)                                317
         END IF                                                   318
         T=T+H                                                    319

*       (Draw curve element)

         GOTO 10                                                  320
        ENDIF                                                     321

*       (Finish drawing; ask whether computation should continue)

        END                                                       322
```

Fig. 5.7. Numerical portion of the main program KAP5

5.4.3 Switch on the driving force, and study the resonance effects in harmonic and anharmonic oscillations.

5.5 Solutions to the Exercises

5.5.1 It is surprising that the Runge-Kutta method with 5 mesh points per oscillation period already gives a profile similar to the true solution curve for the harmonic oscillation. The error in this case is essentially to be found in the damping: whereas the true solution without friction and without driving force is undamped, the approximate solution shows damping. With 10 mesh points per oscillation period, scarcely any visible error remains in the graphical output.

For anharmonic oscillations with $B>2$ the rapid variations in the curvature of the solution curve impose high demands on the solution method. With 10 mesh points per oscillation period the inaccuracy for $B=5$ is still easily detected. Increasing the number of mesh points by a factor of 5 or 10, however, leads even here to a satisfactory accuracy for the graphical output.

Figure 5.8 shows a harmonic oscillation calculated by the Runge-Kutta method with only 5 mesh points per oscillation period.

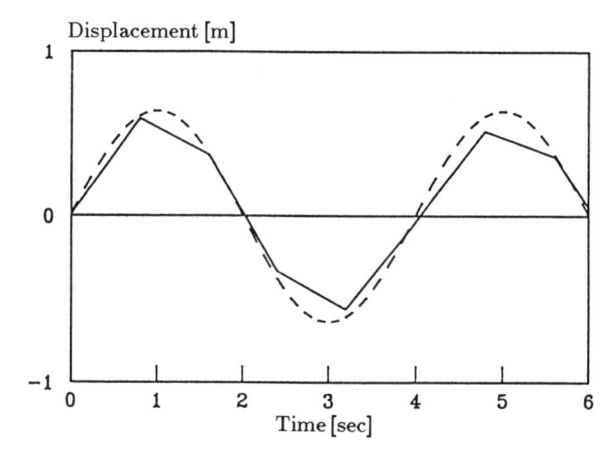

Fig. 5.8. Harmonic oscillation calculated by the Runge-Kutta method using only 5 mesh points per oscillation period; the broken line shows the true solution for comparison

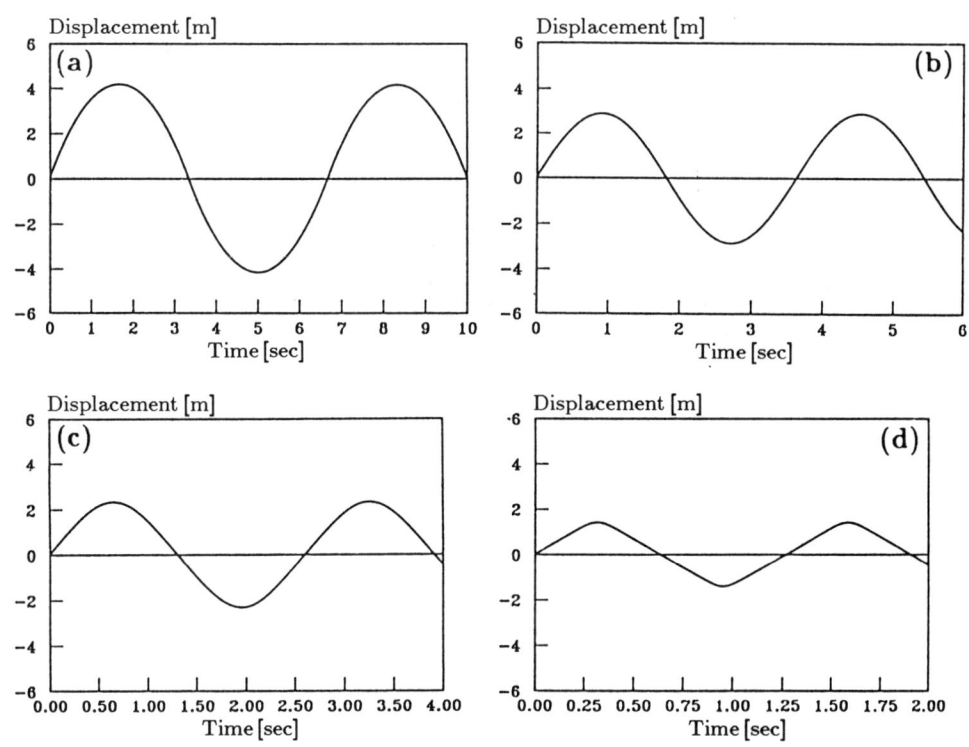

Fig. 5.9a-d. Anharmonic and harmonic oscillations:

a) anharmonic oscillation with $B = 0.00001$,
b) harmonic oscillation,
c) anharmonic oscillation with $B = 2$,
d) strongly anharmonic oscillation with $B = 10$

5.5.2 Figure 5.9 shows oscillation curves for $B = 0.00001, B = 1, B = 2$ and $B = 10$. When $B = 0.00001$ the restoring force is, apart from a change of sign at $x = 0$, independent of the displacement (the value $B = 0.00001$ stands for the value $B = 0$, which is not allowed by the program). The oscillation curve accordingly consists of a sequence of parabolae. For comparison the harmonic oscillation ($B = 1$) is shown. When $B = 2$ the point mass moves in a cubic potential. At the turning points the restoring force is stronger than in the harmonic oscillation. The reversal of motion is accordingly sharper. This effect is shown still more strongly in the case $B = 10$. Here the potential climbs so steeply that the reversal of motion assumes the character of an elastic reflection. In the remaining part of the motion the forces are then comparatively small and have only a slight effect on the velocity of the point mass.

5.5.3 The frequency of the free harmonic oscillation is $\omega = \sqrt{(A/M)}$. If one uses this value in the driving force one has the appropriate starting condition for an oscillation with increasing amplitude (resonance). As friction is ignored in our calculation the amplitude of the oscillation will increase without limit.

When the frequency of the driving force differs slightly from the resonance frequency, e.g. by about 10%, then a beat occurs: depending on the phase difference between the oscillation and the driving force, energy is fed into, or taken from, the oscillation; see for example Fig. 5.10.

With a potential which is only slightly anharmonic (e.g. $B = 1.1$) one can also obtain beats. In this case the amplitude of the oscillation cannot increase without limit, as the anharmonicity ensures that the driving force and the velocity are from time to time in opposition.

With stronger anharmonicity a pronounced beat no longer occurs, because the driving force and the oscillation fall so quickly out of phase. An extreme case is shown in Fig. 5.11. A strong driving force pushes the point mass to and fro. In its motion it occasionally runs into the potential wall ($B = 5$) and gets reflected.

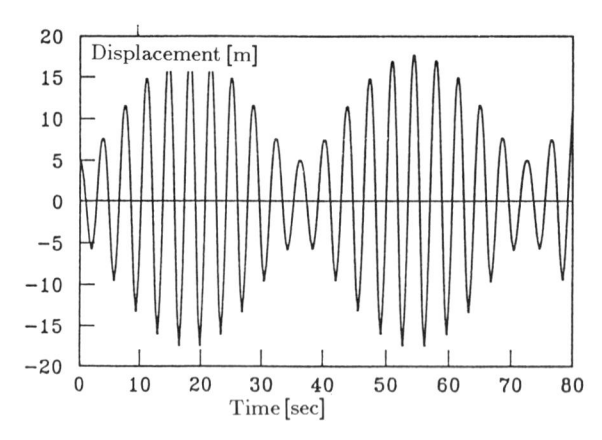

Fig. 5.10. Forced harmonic oscillation. The frequency of the driving force is about 10% greater than the frequency of the free oscillation

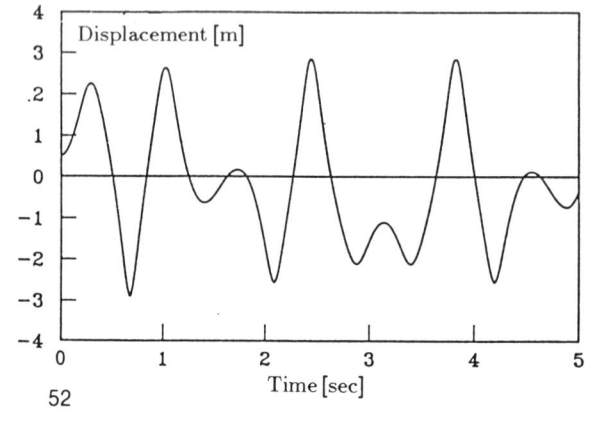

Fig. 5.11. Point mass under the influence of a strong driving force and a strongly anharmonic ($B = 5$) restoring force

6. Coupled Harmonic Oscillations

6.1 Formulation of the Problem

Coupled oscillations occur in many regions of physics. The Raman and infrared spectra, for example, have their origin in the coupled oscillations of atoms within the molecule. The analysis of these oscillations gives information not only on the structure of the molecule but also on the binding forces. Coupled oscillations occur in technology when machine components run roughly. In the design of anchorages, resonance frequencies as well as damping play an important role.

In this chapter we consider as a simple example the coupled harmonic oscillation of two point masses, each weighing 1 kg. They are linked to their rest positions by harmonic restoring forces. The displacements from the rest positions are denoted by x_1 and x_2. If the two displacements are not equal, an additional harmonic force appears, which couples the two bodies to one another. The equation of motion is the following coupled system of differential equations:

$$-M\frac{d^2 x_1}{dt^2} - C_1 x_1 + C(x_2 - x_1) = 0 \quad ,$$
$$-M\frac{d^2 x_2}{dt^2} - C_2 x_2 - C(x_2 - x_1) = 0 \quad . \tag{6.1}$$

In particular, these equations describe the motion of sympathetic pendulums, see Fig. 6.1. When the displacements are small one has approximately harmonic restoring forces. The force constants C_1 and C_2 are determined by the lengths of the pendulums l_i and the force of gravity Mg. The coupling spring has the force constant C. If the pendulum lengths are not equal, then $C_1 \neq C_2$, i.e. the sympathetic pendulums are "out of tune".

With the abbreviations

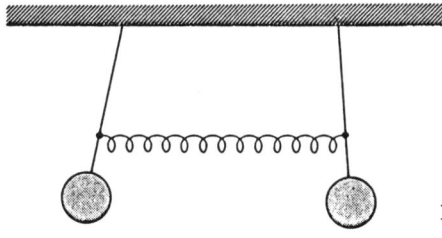

Fig. 6.1. Sympathetic pendulums

$$A_{11} = -(C_1 + C)/M \;, \quad A_{12} = C/M \;,$$
$$A_{21} = C/M \;, \quad\quad\quad A_{22} = -(C_2 + C)/M \tag{6.2}$$

(6.1) can be brought into the general form

$$\frac{d^2 x_1}{dt^2} = A_{11} x_1 + A_{12} x_2 \;, \quad \frac{d^2 x_2}{dt^2} = A_{21} x_1 + A_{22} x_2 \;. \tag{6.3}$$

Generalisation to m bodies gives

$$\frac{d^2 x_i}{dt^2} = \sum_{j=1}^{m} A_{ij} x_j \;, \quad i = 1, \ldots, m \;. \tag{6.4}$$

If necessary frictional and driving forces can be added on the right hand sides.

We have at the outset mentioned technical applications. We recommend to our readers an interesting example for private study. It is the "earthquake-proof skyscraper". In earthquake areas such as, e.g. Tokyo, it has been discovered that skyscrapers should be built not too rigid, if they are to withstand severe earthquakes. It is better to build structures capable of oscillation, and to provide for adequate damping. By the methods which we have now acquired, one can simulate on the computer the behaviour of a multi-storeyed building during an earthquake. One makes the approximate assumption that the whole mass of the building is concentrated in the ceilings, see Fig. 6.2. The steel framework of the building furnishes the restoring forces, when ceilings of successive storeys undergo different lateral displacements. The earthquake is simulated by causing the ground floor to move to and fro in a stipulated manner. The program developed in this chapter can be extended with little effort to calculate the coupled motions of the ceilings and show the solutions $x_i(t)$ on the screen. It will be observed that the building can undergo dangerous characteristic "eigen-oscillations" if no damping is present. With appropriate oscillation damping built in, however, it can be shown on the screen that a skyscraper can well withstand moderately strong earthquakes.

6.2 Numerical Method

As already mentioned in Sect. 4.2.1, a coupled system of second order differential equations can be reduced to a coupled system of first order differential

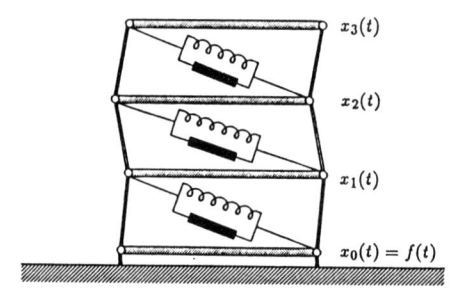

Fig. 6.2. Simulation of the oscillation of a building during an earthquake: the mass is concentrated in the ceilings, restoring and frictional forces act diagonally across the storeys, the ground floor is oscillated to and fro by the earthquake according to the function $f(t)$

equations. Using the notation

$$y_i(t) = x_i(t) , \qquad i = 1,\ldots,n' ,$$
$$y_{i+n'}(t) = \frac{dx_i(t)}{dt} , \qquad i = 1,\ldots,n' , \qquad (6.5)$$

one obtains from (6.4) the system of equations ($n = 2n'$)

$$\frac{dy_i(t)}{dt} = f_i(y_1(t),\ldots,y_n(t);t), \qquad i = 1,\ldots,n , \qquad (6.6)$$

with

$$f_i = y_{n'+i}(t) , \qquad f_{n'+i} = \sum_{j=1}^{n'} A_{ij} y_j(t) , \qquad i = 1,\ldots,n' . \qquad (6.7)$$

In order to solve the system of differential equations (6.6), we shall employ the Runge-Kutta method presented in Sect. 5.2.2.

6.3 Programming

The functions f_i on the right-hand sides of our system of equations (6.6) are, according to (6.7) and (6.2),

$$f_1 = y_3(t) , \qquad (6.8a)$$

$$f_2 = y_4(t) , \qquad (6.8b)$$

$$f_3 = -\left(\frac{C_1}{M} + \frac{C}{M}\right) y_1(t) + \frac{C}{M} y_2(t) , \qquad (6.8c)$$

$$f_4 = \frac{C}{M} y_1(t) - \left(\frac{C_2}{M} + \frac{C}{M}\right) y_2(t) . \qquad (6.8d)$$

The subroutine DGL6 (Figs. 6.3 and 6.4) calculates the functions (6.8a-d) in lines 105 to 108. The dimensions of the arrays Y and F must be raised from 2 to 4, since we now have four differential equations (line 104).

For the solution of the system (6.6) we apply the subroutine RUNGE from the previous chapter. The main program KAP6 is similar to the program KAP5 (see Fig. 6.6). In the dimensioning of Y (line 203) and in the call for RUNGE (line 214) we have to take note that we now have to solve four coupled equations.

Physical title	FORTRAN title	Physical title	FORTRAN title
C_1	C1	t	T
C_2	C2	$y_i(t)$	Y(I)
C	C	$f_i(y_1(t), y_2(t),$	
M	M	$y_3(t), y_4(t); t)$	F(I)

Fig. 6.3. Notation employed in subroutine DGL6

```
      SUBROUTINE DGL6(T,Y,F)                                100

      IMPLICIT DOUBLE PRECISION (A-H,O-Z)                   101
      DOUBLE PRECISION M                                    102
      COMMON /DAT/ C1, C2, C, M                             103
      DIMENSION Y(4), F(4)                                  104
      F(1)=Y(3)                                             105
      F(2)=Y(4)                                             106
      F(3)=-C1/M*Y(1)+C/M*(Y(2)-Y(1))                       107
      F(4)=-C2/M*Y(2)-C/M*(Y(2)-Y(1))                       108
      END                                                   109
```

Fig. 6.4. Subroutine DGL6

Physical title	FORTRAN title	Physical title	FORTRAN title
C_1	C1	$y_1(t) = x_1(t)$	Y(1)
C_2	C2	$y_2(t) = x_2(t)$	Y(2)
C	C	$y_3(t) = dx_1/dt$	Y(3)
M	M	$y_4(t) = dx_2/dt$	Y(4)
$y_1(0)$	X10	h	H
$y_2(0)$	X20	t	T
$y_3(0)$	V10	t_1	T1
$y_4(0)$	V20		

Fig. 6.5. Notation employed in main program KAP6

```
      PROGRAM KAP6                                          200

      IMPLICIT DOUBLE PRECISION (A-H,O-Z)                   201
      DOUBLE PRECISION M                                    202
      DIMENSION Y(4)                                        203
      EXTERNAL DGL6                                         204
      COMMON /DAT/ C1, C2, C, M                             205

*     (Input of C1,C2,C,X10,X20,V10,V20,T1,H)

      M=1.D0                                                206
      T=0.D0                                                207
      Y(1)=X10                                              208
      Y(2)=X20                                              209
      Y(3)=V10                                              210
      Y(4)=V20                                              211

10    CONTINUE                                              212
      IF(T.LT.T1) THEN                                      213
        CALL RUNGE(4,H,T,Y,DGL6)                            214
        T=T+H                                               215

*       (graphical output of T,X1,X2)

        GOTO 10                                             216
      ENDIF                                                 217

      END                                                   218
```

Fig. 6.6. Main program KAP6

We therefore need four starting conditions, namely the displacements and the velocities of the two bodies at time $t = 0$.

In the graphical output we make use of the possibility of drawing several curves at once. To achieve this, for each call of the subroutines GMOVE and GDRAW we assign the parameter NUMBER the value 1 for one curve, and the value 2 for the other curve.

The complete program is stored on the diskette in the file KAP-06.FOR

6.4 Exercises

6.4.1 Study the nature of the oscillations of sympathetic pendulums ($C_1 = C_2$) with weak coupling ($C = C_1/10$) and with strong coupling ($C = 10C_1$).

6.4.2 Choose a moderately strong coupling parameter C and unequal force constants C_1, C_2. Study the energy transfer between the two point masses for various initial conditions. Are there initial conditions for which no transfer of energy takes place?

6.5 Solutions to the Exercises

6.5.1 With weak coupling as a rule one pendulum transfers energy to the other pendulum until it no longer has any left. Then the process is reversed, see Fig. 6.7. One should notice the relative phases of the oscillation curves. The pendulum which is delivering the energy has a phase lead of up to 90°. During the course of the energy transfer the phase lead is decreased and eventually becomes negative, when the driving pendulum becomes the driven. No transfer of energy takes place if the two pendulums are oscillating with equal amplitude either in phase or 180° out of phase.

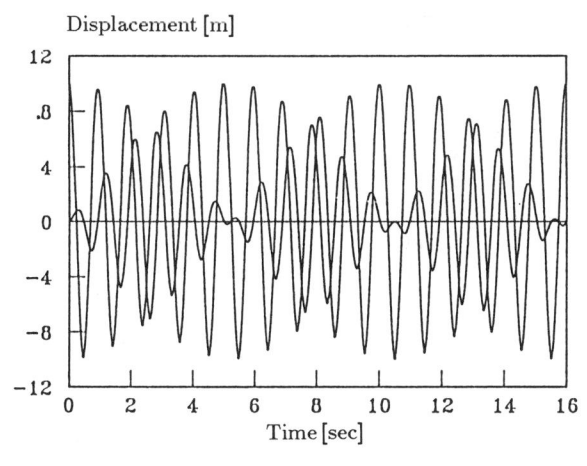

Fig. 6.7. Oscillations of sympathetic pendulums with weak coupling ($C_1 = C_2 = 39.5\,\mathrm{Nm}^{-1}, C = 3.95\,\mathrm{Nm}^{-1}$)

With strong coupling the two pendulums can oscillate relative to one another with a short oscillation period, whilst their combined centre of mass executes a pendulum motion with longer oscillation period, see Fig. 6.8.

6.5.2 With unequal force constants C_1 and C_2 (unequal pendulum lengths) the energy is now completely transferred only in one direction, see Fig. 6.9. In the other direction the transfer of energy is only partial.

Fig. 6.8. Oscillations of sympathetic pendulums with strong coupling ($C_1 = C_2 = 39.5\,\mathrm{Nm}^{-1}, C = 395\,\mathrm{Nm}^{-1}$)

Fig. 6.9. Coupled oscillation with $C_1 = 40, C_2 = 30$ and $C = 10\,\mathrm{Nm}^{-1}$

Fig. 6.10. One of the two eigen-oscillations of the system of two point masses; the force constants are the same as in Fig. 6.9

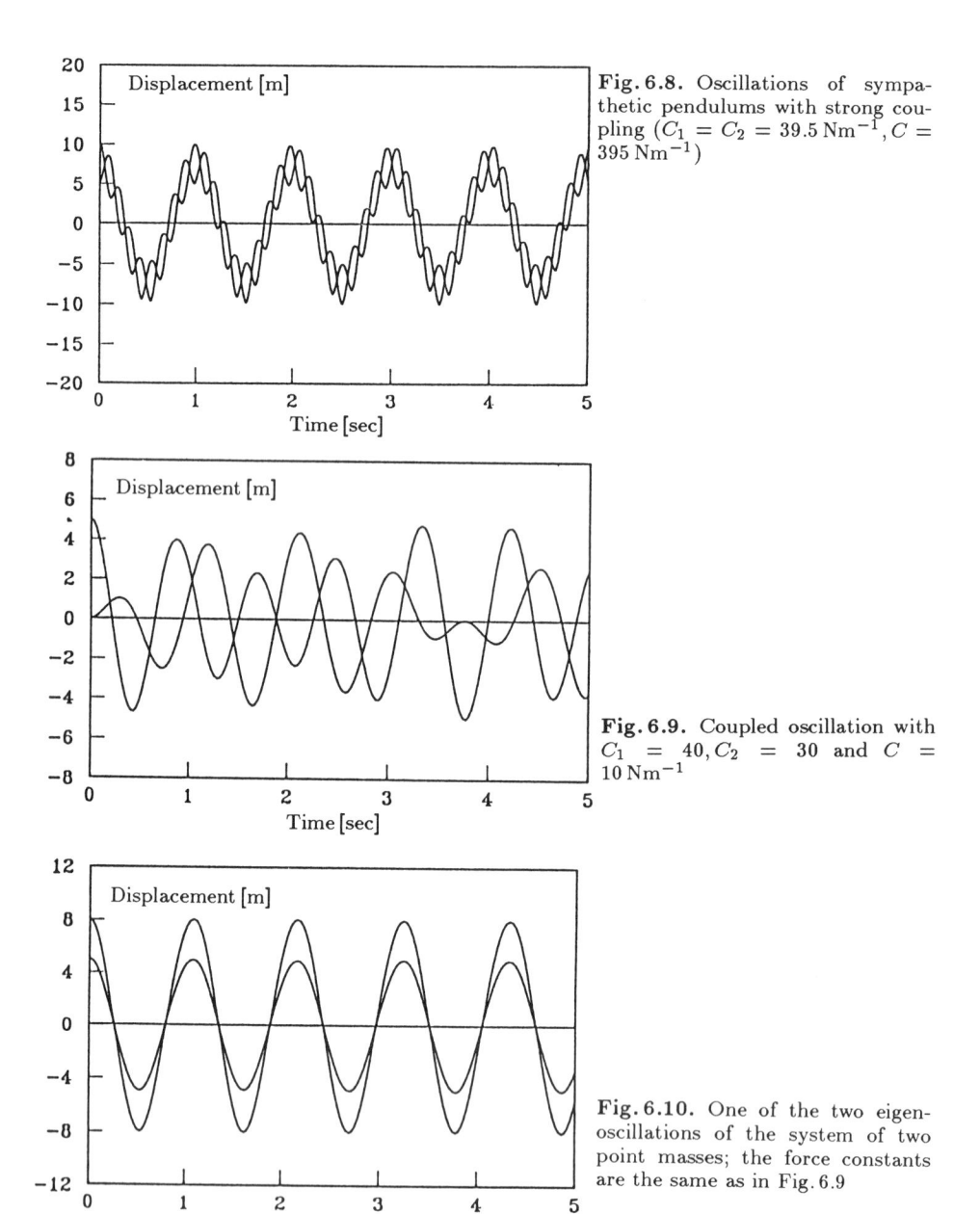

There are initial conditions for which no energy transfer takes place. Fig. 6.10 shows an oscillation with the same force constants as in Fig. 6.9, but with the initial conditions $y_1(0) = 5\mathrm{m}$, $y_2(0) = 8\mathrm{m}$, $y_3(0) = y_4(0) = 0$. Such an oscillation is called an eigen-oscillation of the system. In our example there are two eigen-oscillations. Their oscillation periods are about 1.08 and 0.84 s. The search for the initial conditions leading to eigen-oscillations is equivalent to the search for so-called normal coordinates. The latter are obtained by an orthogonal tranformation from the coordinates used by us. The transformation to normal coordinates causes the non-diagonal elements of the matrix (A_{ij}) of the equations (6.4) to vanish.

7. The Flight Path of a Space Craft as a Solution of the Hamilton Equations

7.1 Formulation of the Problem

In textbooks and lectures on mechanics, after Newton's equations of motion it is usual to consider the Lagrange equations of the first and second sort and thereafter Hamilton's equations. One first of all gets the impression that with the constant introduction of new equations mechanics is becoming not simpler, but more complicated. Only by an attempt to solve a non-trivial problem of particle mechanics — such as the simultaneous motion of two planets round the sun — does one come to realise that Hamilton's equations are helpful.

In this chapter we shall consider a somewhat complicated example of particle mechanics with the aid of Hamilton's equations. For this we choose the flight path of a space craft from the earth to the moon and back. We shall see that it is not easy to choose the starting conditions so that the space craft comes into the general vicinity of the moon. We shall see that it is much more difficult still to find a path in which the space craft is swung round by the gravitation of the moon so that it returns again to the vicinity of the earth. The astronauts of "Apollo 13", whose engine failed on the flight to the moon, had to follow such a flight path back to earth. In order to make the situation realistic, we shall also provide the possibility of a "braking force" *en route,* in order, for example, to bring the space craft into an orbit round the moon.

We shall limit ourselves to a simplified model of earth, moon and space craft, and ignore many refinements such as, for example, the influence of the sun and the other planets. Within the framework of this model, however, we shall calculate accurately. The Euler method is unsuitable here, and we have to improve even the Runge-Kutta method by introducing a variable mesh width.

Compared with Newton's equations of motion, Hamilton's equations have two advantages:

1. They are differential equations of the first order. The formation of differences of the second order (see Chap. 2) is thereby henceforth avoided.
2. Hamilton's equations can be written down in any coordinate system appropriate for the problem.

Our recasting of the oscillation equations in Chap. 4 by means of $y_1 = x$, $y_2 = dx/dt$ was already the transformation to the Hamilton equations. In this recasting we have doubled the number of unknown functions and instead of a differential equation of the second order we have obtained two differential equations of the first order. Since the problem was very simple it was not noticed that we were already working with Hamilton's equations.

In this book we shall not derive Hamilton's equations. The derivation is to be found in text-books on classical mechanics. We shall, however, show how the equations are set up and solved in practice.

It is not difficult to formulate Hamilton's equations for a system of point masses, between which there are forces generated by a potential. One first chooses a convenient system of coordinates $q_k(t)$, $k = 1, \ldots, n'$, to describe the positions of the point masses as functions of the time t. The unknown functions in Hamilton's equations are the position coordinates $q_k(t)$ and the momentum coordinates $p_k(t)$ which are canonically conjugate to them. To determine the momentum coordinates one forms the Lagrange function,

$$L = T\left(q_k, \frac{dq_k}{dt}\right) - V\left(q_k, \frac{dq_k}{dt}, t\right) , \tag{7.1}$$

where T is the kinetic and V the potential energy. The potential V may depend explicitly on the time. The canonical momentum coordinates p_k are defined as the partial derivatives

$$p_k = \frac{\partial L}{\partial (dq_k/dt)} . \tag{7.2}$$

One now expresses the Hamiltonian function H' as a function of the coordinates p_k, q_k and dq_k/dt. It is defined as

$$H' = \sum_k p_k \frac{dq_k}{dt} - L . \tag{7.3}$$

With (7.2) one has a relation between p_k, q_k and dq_k/dt. One makes use of this relation in order to eliminate the coordinates dq_k/dt from the Hamiltonian function H' and express it in terms of the canonical coordinates q_k and p_k. We denote this Hamiltonian function in the canonical coordinates by H,

$$H = H(q_k, p_k) . \tag{7.4}$$

Hamilton's equations then become

$$\frac{dq_k}{dt} = \frac{\partial H}{\partial p_k} , \quad k = 1, \ldots, n' ,$$
$$\frac{dp_k}{dt} = -\frac{\partial H}{\partial q_k} , \quad k = 1, \ldots, n' . \tag{7.5}$$

Hamilton's equations are a system of differential equations of the first order. Having specified the initial values $q_k(t_0)$, $p_k(t_0)$ one solves the equations to obtain $q_k(t)$ and $p_k(t)$.

For the remaining mathematical treatment it is convenient to rename the coordinates,

$$(q_1, \ldots, q_{n'}, p_1, \ldots, p_{n'}) \equiv (y_1, \ldots, y_n) , \quad n = 2n' , \tag{7.6}$$

and write Hamilton's equations (7.5) in the form

$$\frac{dy_i}{dt} = f_i(y_1, \ldots, y_n; t) \quad , \quad i = 1, \ldots, n \quad . \tag{7.7}$$

We come now to our practical example. As a simplification we shall assume that the moon moves on a circular orbit round the earth, with an orbital period of 27 days = 648 hours. We shall assume the earth-moon distance to be 384 000 km, corresponding to the mean distance between the two bodies. The computer could carry out the computation in Cartesian coordinates. However, we shall make use of the freedom of choice of coordinates, and compute in plane polar coordinates. The origin of coordinates is the centre of mass of the earth-moon system. The influence of the sun is neglected. The mass of the moon is in fact about 1/81 of the mass of the earth. In order to "hit" the moon more easily at first, we shall make the moon mass an input parameter, and set it, for example, equal to 1/20 or 1/50 of the earth mass. We shall identify quantities pertaining to earth by the index 1, and those pertaining to the moon by the index 2. At the time $t = 0$ the earth and the moon have the coordinates

$$r_1 = \frac{m_2}{m_1 + m_2} 384\,000\,\text{km} \quad , \quad \varphi_1 = 180° \quad ,$$

$$r_2 = \frac{m_1}{m_1 + m_2} 384\,000\,\text{km} \quad , \quad \varphi_2 = 0° \quad . \tag{7.8}$$

The mass μ of the space craft is vanishingly small compared with the masses of the moon and the earth, i.e. the motion of the space craft has no effect on the motions of moon and earth. Our example is therefore not a true three-body problem, but only the problem of a body moving in a gravitational potential with a specified dependency on time.

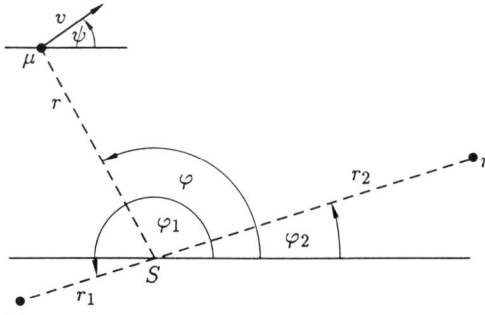

Fig. 7.1. Earth, moon, space craft and their polar coordinates in the centre of mass system

The solution we seek is the position r of the space craft as a function of time,

$$\boldsymbol{r}(t) = (r(t), \varphi(t)) \quad . \tag{7.9}$$

We also find the velocity $\boldsymbol{v}$ of the space craft in polar coordinates:

$$\boldsymbol{v}(t) = \left(\left| \frac{d\boldsymbol{r}(t)}{dt} \right|, \psi(t) \right) \quad , \tag{7.10}$$

where $\psi(t)$ is the polar angle indicating the direction of the velocity. The duration of time, during which the motor of the space craft gives thrust, is in practice small compared with the total time of flight. We let this time go to zero and say: immediately after the start the space craft has the initial position $r(0)$, $\varphi(0)$ and the initial velocity $v(0)$, $\psi(0)$. At any intermediate specified time t we can switch on the motor again and thereby cause the space craft to attain a new, specified, velocity. Between the times 0 and t Hamilton's equations determine the path of the space craft.

For setting up the Lagrange function we need the distances of the space craft from the earth and the moon,

$$s_1 = \sqrt{r^2 + r_1^2 - 2rr_1 \cos(\varphi - \varphi_1)}\;, \tag{7.11a}$$

$$s_2 = \sqrt{r^2 + r_2^2 - 2rr_2 \cos(\varphi - \varphi_2)}\;. \tag{7.11b}$$

The Lagrange function is then

$$\begin{aligned}L &= T - V \\ &= \frac{\mu}{2}\left(\left(\frac{dr}{dt}\right)^2 + r^2\left(\frac{d\varphi}{dt}\right)^2\right) + \frac{\gamma m_1 \mu}{s_1} + \frac{\gamma m_2 \mu}{s_2}\;.\end{aligned} \tag{7.12}$$

The momentum coordinates canonically conjugate to r and φ are, from (7.2),

$$p_r = \mu \frac{dr}{dt}\;, \tag{7.13a}$$

$$p_\varphi = \mu r^2 \frac{d\varphi}{dt}\;. \tag{7.13b}$$

From (7.3) it then follows that

$$\begin{aligned}H' &= \mu \frac{dr}{dt}\frac{dr}{dt} + \mu r^2 \frac{d\varphi}{dt}\frac{d\varphi}{dt} - L \\ &= \frac{\mu}{2}\left(\left(\frac{dr}{dt}\right)^2 + r^2\left(\frac{d\varphi}{dt}\right)^2\right) - \frac{\gamma m_1 \mu}{s_1} - \frac{\gamma m_2 \mu}{s_2}\;.\end{aligned} \tag{7.14}$$

We have to express dr/dt and $d\varphi/dt$ in terms of p_r and p_φ, and so obtain

$$H = \frac{p_r^2}{2\mu} + \frac{p_\varphi^2}{2\mu r^2} - \frac{\gamma m_1 \mu}{s_1} - \frac{\gamma m_2 \mu}{s_2}\;. \tag{7.15}$$

Taking account of (7.11), Hamilton's equations (7.5) then become

$$\frac{dr}{dt} = \frac{p_r}{\mu}\;, \tag{7.16a}$$

$$\frac{d\varphi}{dt} = \frac{p_\varphi}{\mu r^2}\;, \tag{7.16b}$$

$$\frac{dp_r}{dt} = \frac{p_\varphi^2}{\mu r^3} - \frac{\gamma m_1 \mu}{s_1^3}[r - r_1 \cos(\varphi - \varphi_1)]$$
$$- \frac{\gamma m_2 \mu}{s_2^3}[r - r_2 \cos(\varphi - \varphi_2)] \quad , \tag{7.16c}$$

$$\frac{dp_\varphi}{dt} = -\frac{\gamma m_1 \mu}{s_1^3} r r_1 \sin(\varphi - \varphi_1) - \frac{\gamma m_2 \mu}{s_2^3} r r_2 \sin(\varphi - \varphi_2) \quad . \tag{7.16d}$$

The gravitational forces contained in the last two equations are explicitly time dependent, since φ_1 and φ_2 are functions of time:

$$\varphi_1 = \omega t \quad , \quad \varphi_2 = \omega t + \pi \quad , \tag{7.17}$$

where ω denotes the angular velocity with which the earth and the moon move round their combined centre of mass.

The coordinate system used by us for the computation, whose origin is the total centre of mass of the earth and the moon, is not very convenient for the input and output. For a space craft, quantities such as distance and velocity relative to the earth or the moon are more interesting than the distance to the total centre of mass of the system. We therefore want to provide in the program the capability of expressing the input and output in, altogether, 6 different coordinate systems. The systems 0 to 2 use as origin the centre of mass of, respectively, the earth-moon system (0), the earth (1) or the moon (2). In system 0 we solve Hamilton's equations. In system 1 or 2 one can answer questions such as "What is the least velocity relative to the earth that a rocket must achieve, in order to fly from the earth to the moon?" or: "How much energy is needed to take a space craft from the lunar surface into a lunar orbit?" The systems 3 to 5 have the same origins as 0 to 2, but rotate with the line joining earth and moon, so that these celestial bodies are at rest in the coordinate systems. The observer is then apparently sitting on a beam joining earth and moon. In the coordinate systems 3 to 5 it is especially easy to study the long-term perturbations caused by the moon on terrestrial satellites in wide orbits.

7.2 Mathematical Methods

7.2.1 Mesh Width Adaptation in the Runge-Kutta Method

In the Runge-Kutta method we have a very accurate method for the solution of differential equations, which we can also use in Hamilton's equations (7.16a–d). The problem here, of course, is the choice of a suitable mesh width, since strength and direction of gravitation are strongly dependent on position.

A space craft in a close terrestrial orbit has an orbital period of about 90 minutes, i.e. the direction of the velocity and the direction of the gravitational force change through 360° in this time. The temporal mesh width should therefore be in the order of minutes. On the other hand, a space craft which finds itself far from the earth and the moon, for example, half-way between the two, changes its velocity only slightly over a period of hours. If one works here with a mesh width in the order of minutes the program will be slow. For at every evaluation of (7.16c) and (7.16d) functions have to be computed, which take a lot of computing time. Moreover a mesh width which is too small raises the danger that rounding errors will accumulate.

During the motion of the space craft we shall therefore adjust the mesh width to the current circumstances. A simple algorithm will carry out this adjustment automatically.

To demonstrate the automatic adjustment of the mesh width it is sufficient to discuss the method for a one-dimensional equation

$$\frac{dy}{dt} = f(y,t) \quad . \tag{7.18}$$

We start from the situation that we know the initial value $y(t_0)$ of the solution function and wish to determine a value $y(t_1)$. In our example t_0 will be the instant at which we have plotted the latest position of the space craft, and t_1 the time for the next position to be determined. We suppose that nowhere on the portion of the flight path between t_0 and t_1 will the function $f(y,t)$ become more than a few powers of ten greater than at the end-points. We denote by $y_h(t_1)$ the value of the solution function at the point t_1, obtained by integration of (7.18) using the Runge-Kutta method with a tentatively chosen mesh width h. We denote the amount of the deviation from the true value by ε_h,

$$\varepsilon_h = |y_h(t_1) - y(t_1)| \quad . \tag{7.19}$$

For each individual step the Runge-Kutta method is of order h^5, according to Sect. 5.2. For sufficiently small h, therefore, the error per step decreases by the factor 1/32 when the step is halved. On the other hand, when the step is halved we need twice as many steps. Accordingly we find that:

$$\varepsilon_h \approx 16\varepsilon_{h/2} \quad , \tag{7.20}$$

and, as a result,

$$\varepsilon_h \approx \tfrac{16}{15}(\varepsilon_h - \varepsilon_{h/2}) = \tfrac{16}{15}|(y_h(t_1) - y_{h/2}(t_1))| \quad . \tag{7.21}$$

The integration step, or mesh width h, can be adjusted in the following way:

1. We specify a limit $\varepsilon_{\max}$ and seek to choose h so that the error for an individual integration step is not greater than $\varepsilon_{\max}$. First of all we try integrating from t_0 to t_1 in one step. Accordingly we set

$$h = t_1 - t_0 \quad . \tag{7.22a}$$

In what follows we denote the starting point for an individual Runge-Kutta step by t. Initially we therefore have:

$$t = t_0 \ . \tag{7.22b}$$

2. We integrate (7.18) with the Runge-Kutta method from t to $t+h$ using mesh widths h and $h/2$. We compare the two function values so obtained and from (7.21) obtain an estimate for ε_h.

3. Since the Runge-Kutta method for an individual step is of order h^5, we can estimate the maximal value allowed for the mesh width:

$$\left|\frac{\varepsilon_{\max}}{\varepsilon_h}\right| = \left(\frac{h_{\max}}{h}\right)^5 \ . \tag{7.23}$$

4. Now we consider two cases:

   a) If $h_{\max} < h/2$, we replace h by $2h_{\max}$ and go back to stage 2.

   b) If $h_{\max} \geq h/2$, we accept $y_{h/2}(t+h)$ as the valid result for $y(t+h)$. If h has not been changed, then $t+h$ is equal to t_1, and we have achieved the integration goal t_1. If we have changed h, then it is possible that $t+h$ has still not reached t_1. In this case we set t equal to $t+h$ and go back to stage 2. If the new integration step takes us past the integration goal t_1, then h is decreased so that $t+h$ is just equal to t_1.

5. In the programming of the mesh width adjustment we take account of the following special features:

   a) We specify a lower limit for the mesh width h, below which the program aborts with an error message.

   b) At each integration step we test whether h can be increased again for the next step. This protects us against continuing up to the time t_1 with an unnecessarily small time step, when h has become very small between t_0 and t_1.

The generalisation to several coupled equations is straightforward. For each dependent variable y_i we obtain an estimate for the error from (7.21). We denote the largest estimate by ε_h and proceed with this as described above. The choice of an unique error limit for all y_i is possible only when the latter do not differ by too many orders of magnitude. If our differential equation describes a physical process, we can usually achieve this by choice of suitable units.

The mesh width adjustment described above does not assume that t_1 is greater than t_0. The mesh width h may be negative, without invalidating any of the equations (7.19–23). This makes possible an additional test of accuracy, by "integrating backwards", i.e. computing $y(t_0)$ from $y(t_1)$. If the $y(t_0)$ determined by computing backwards differs too much from the initial value, we know that $\varepsilon_{\max}$ has been set too high.

There are also methods of mesh width adjustment in the Runge-Kutta method which use the intermediate values of $k^{(1)}$ to $k^{(4)}$ in order to estimate the discretisation error. With these methods one does not need to carry out each integration step several times. The methodology, however, is tuned to the Runge-Kutta method and cannot be transferred to other integration methods without further work. The method given above, on the other hand, can very easily be generalised and is therefore frequently used.

7.2.2 Coordinate Transformation

The coordinate systems 1 to 5, which we use for input and output, differ from the coordinate system 0, in which we solve Hamilton's equations, by a (time-dependent) rotation and translation:

$$r(t) = r_c(t) + O(t) r_d(t) \quad . \tag{7.24}$$

We denote the position vector in the coordinate system 0 by $r(t)$, those in the coordinate systems 1 to 5 by $r_d(t)$. The vector $r_c(t)$ describes the shift in the origin of coordinates, the matrix $O(t)$ describes the rotation. In the coordinate system 3 r_c is zero, whilst in the coordinate systems 1 and 2 the rotation angle is zero. If we set r_c and the rotation angle to zero, (7.24) describes the identical transformation. Accordingly we can use the subroutine which we shall write for the evaluation of (7.24) even if we carry out the input and output in the coordinate system 0.

The addition of vectors can be carried out more easily in Cartesian coordinates than in polar coordinates. Accordingly in the coordinate transformation we compute with Cartesian coordinates. We have:

$$r_c(t) = (x_c(t), y_c(t)) = (r_c \cos \omega t, r_c \sin \omega t) \quad . \tag{7.25}$$

For the coordinate systems 0 and 3 r_c is equal to zero, for 1 and 4 equal to $-r_1$, for 2 and 5 equal to r_2. The minus sign in front of r_1 takes account of the fact that the earth's position vector (from the total centre of mass) is at an angle π relative to that of the moon.

The rotation matrix $O(t)$ is given by

$$O(t) = \begin{pmatrix} \cos \omega_r t & -\sin \omega_r t \\ \sin \omega_r t & \cos \omega_r t \end{pmatrix} \quad . \tag{7.26}$$

The angular velocity ω_r, with which the coordinate system rotates, is equal to ω for the coordinate systems 3 to 5, and 0 for the non-rotating systems 0 to 2.

The transformation equation for the velocity is obtained by differentiating both sides of (7.24) with respect to the time. One gets

$$\begin{aligned} \frac{dr}{dt} &= \frac{dr_c}{dt} + O \frac{dr_d}{dt} + \frac{dO}{dt} r_d \\ &= \frac{dr_c}{dt} + O \left(\frac{dr_d}{dt} + O^{-1} \frac{dO}{dt} r_d \right) \quad . \end{aligned} \tag{7.27}$$

From (7.25) one obtains

$$\frac{d\boldsymbol{r}_c}{dt} = (-\omega r_c \sin \omega t, \omega r_c \cos \omega t) = (-\omega y_c, \omega x_c) \;, \tag{7.28}$$

and from (7.26)

$$\frac{dO}{dt} = \begin{pmatrix} -\omega_r \sin \omega_r t & -\omega_r \cos \omega_r t \\ \omega_r \cos \omega_r t & -\omega_r \sin \omega_r t \end{pmatrix} \;. \tag{7.29a}$$

Since O is a rotation matrix, its inverse is equal to its transpose, and we obtain by multiplying out

$$O^{-1} \frac{dO}{dt} = \begin{pmatrix} 0 & -\omega_r \\ \omega_r & 0 \end{pmatrix} \;, \tag{7.29b}$$

which greatly simplifies the evaluation. The vector $O^{-1}(dO/dt)\boldsymbol{r}_d$ is perpendicular to $\boldsymbol{r}_d$. In what follows we shall denote it by $\boldsymbol{v}_r$, and its components by $v_{r,x}$, $v_{r,y}$.

For Hamilton's equations we need the canonical momenta. We calculate them directly from the Cartesian coordinates for position and velocity. Since the path of a body in a gravitational field does not depend on its mass, we can set the mass μ of the space craft equal to unity and obtain from (7.13):

$$p_r = \frac{dr}{dt} = \frac{d\sqrt{x^2+y^2}}{dt} = \frac{1}{r}\left(x\frac{dx}{dt} + y\frac{dy}{dt}\right) \;, \tag{7.30a}$$

$$p_\varphi = x\frac{dy}{dt} - \frac{dx}{dt}y \;. \tag{7.30b}$$

Equation (7.30b) is most simply obtained by noticing from (7.13b) that p_φ is the angular momentum.

In the transformation from the canonical coordinate system into the input/output system we have to invert the system of equations (7.30). This results in:

$$\frac{dx}{dt} = \frac{r p_r x - y p_\varphi}{r^2} \;, \tag{7.31a}$$

$$\frac{dy}{dt} = \frac{r p_r y + x p_\varphi}{r^2} \;. \tag{7.31b}$$

7.3 Programming

7.3.1 Hamilton's Equations of Motion

In order to bring our system of differential equations (7.16) into the form (7.7),

$$\frac{dy_i}{dt} = f_i(y_1, y_2, y_3, y_4; t) \;, \quad i = 1, \ldots, 4 \;, \tag{7.32}$$

we first have to number our coordinates r, φ, p_r and p_φ according to (7.6). We set

$$y_1 = r, \quad y_2 = \varphi,$$
$$y_3 = p_r, \quad y_4 = p_\varphi. \tag{7.33}$$

The system of equations (7.16), with $\mu = 1$, then becomes:

$$f_1 = y_3, \tag{7.34a}$$

$$f_2 = \frac{y_4}{y_1^2}, \tag{7.34b}$$

$$f_3 = \frac{y_4^2}{y_1^3} - \frac{\gamma m_1}{s_1^3}[y_1 - r_1 \cos(y_2 - \varphi_1)]$$
$$- \frac{\gamma m_2}{s_2^3}[y_1 - r_2 \cos(y_2 - \varphi_2)], \tag{7.34c}$$

$$f_4 = -\frac{\gamma m_1}{s_1^3} y_1 r_1 \sin(y_2 - \varphi_1) - \frac{\gamma m_2}{s_2^3} y_1 r_2 \sin(y_2 - \varphi_2). \tag{7.34d}$$

The equations (7.34a–d) are evaluated in subroutine DGL7. The expressions for f_1 and f_2 can be programmed immediately. The expressions for f_3 and f_4 contain a few transcendental functions, whose computation is time-consuming. Since DGL7 is called four times for each Runge-Kutta recursion step, one should try to economise in computation time by first computing the expressions which are common to (7.34c) and (7.34d). Since the earth is at an angle π in advance of the moon, we have:

$$\varphi_1 = \varphi_2 + \pi, \tag{7.35}$$

$$\sin(y_2 - \varphi_1) = -\sin(y_2 - \varphi_2),$$
$$\cos(y_2 - \varphi_1) = -\cos(y_2 - \varphi_2). \tag{7.36}$$

Of the sine and cosine expressions occurring in (7.34c) and (7.34d) we therefore

Physical title	FORTRAN title	Physical title	FORTRAN title
$y_1 = r$	Y(1)	γm_1	G1
$y_2 = \varphi$	Y(2)	γm_2	G2
$y_3 = p_r$	Y(3)	r_1	R1
$y_4 = p_\varphi$	Y(4)	r_2	R2
ω	OMEGA	s_1	S1
t	T	s_2	S2
$\varphi_2 = \omega t$	PHI2	s_1^3	S13
$\sin(\varphi - \varphi_2)$	SIPH	s_2^3	S23
$\cos(\varphi - \varphi_2)$	COPH	$f_i(y_1(t), y_2(t), y_3(t), y_4(t); t)$	F(I)

Fig. 7.2. Notation employed in subroutine DGL7

need only calculate one of each. For economy of computation time it is also desirable to compute s_1^3 and s_2^3 beforehand.

Figure 7.2 presents the notation used in DGL7. The quantities R1, R2, G1, G2 and OMEGA are needed also in other subroutines. They therefore belong to a COMMON block called /SYST/.

The subroutine DGL7 can be written very compactly with these definitions (see Fig. 7.3).

```
      SUBROUTINE DGL7(T,Y,F)                                   100
      IMPLICIT DOUBLE PRECISION(A-H,O-Z)                       101
      DIMENSION F(4),Y(4)                                      102
      COMMON/SYST/G1,R1,G2,R2,OMEGA,OMEGAR,RCENT,PLOTR,NCOR,PI 103
      F(1)=Y(3)                                                104
      F(2)=Y(4)/Y(1)**2                                        105
      PHI2=OMEGA*T                                             106
      COPH=COS(Y(2)-PHI2)                                      107
      SIPH=SIN(Y(2)-PHI2)                                      108
      S1=SQRT( R1*R1 + Y(1)*Y(1) + 2*R1*Y(1)*COPH )            109
      S2=SQRT( R2*R2 + Y(1)*Y(1) - 2*R2*Y(1)*COPH )            110
      S13=S1**3                                                111
      S23=S2**3                                                112
      F(3)=(Y(4)**2)/(Y(1)**3)                                 113
     &     -G1*(Y(1)+R1*COPH)/S13                              114
     &     -G2*(Y(1)-R2*COPH)/S23                              115
      F(4)= G1*(Y(1)*R1*SIPH)/S13                              116
     &     -G2*(Y(1)*R2*SIPH)/S23                              117
      END                                                      118
```

Fig. 7.3. Subroutine DGL7

7.3.2 Automatic Mesh Width Adjustment in the Runge-Kutta Method

The automatic mesh width adjustment is accomplished in the subroutine FINTEG. Figure 7.4 shows the FORTRAN titles of the quantities used in Sect. 7.2.1. The program FINTEG is listed in Fig. 7.5.

The array Y(I) assumes the actual values $y_1(t)$, $y_2(t)$, ..., $y_n(t)$. When FINTEG is called it holds the values $y_i(t_0)$, at the end it holds the computed values $y_i(t_1)$. For all y_i we use the same error limit ε_{max}. We must therefore

Mathematical title	FORTRAN title	Mathematical title	FORTRAN title
t_0	T0	t	T
t_1	T1	$t+h$	TNEW
ε_h	EPS	h	H
ε_{max}	TOL	h_{max}	HMAX
$y_i(t)$	Y(I)	h/h_{max}	RELA
$y_{i,h}(t)$	Y1(I)		
$y_{i,h/2}(t)$	Y2(I)		

Fig. 7.4. Notation employed in subroutine FINTEG

```
      SUBROUTINE FINTEG(T0,T1,N,Y,TOL,DGL,IERR)                  200

      IMPLICIT DOUBLE PRECISION(A-H,O-Z)                         201
      PARAMETER(NMAX=10)                                         202
      EXTERNAL DGL                                               203
      DIMENSION Y(N)                                             204
      DOUBLE PRECISION Y1(NMAX),Y2(NMAX)                         205

      IF (N.GT.NMAX.OR.N.LT.0.OR.TOL.LE.0)                       206
     &    STOP 'False parameter for FINTEG'                      207
      IERR=0                                                     208
      T   =T0                                                    209
      TNEW=T1                                                    210
      H=T1-T0                                                    211
      IF (H.EQ.0.D0) RETURN                                      212

1     CONTINUE                                                   213
      DO 10 I=1,N                                                214
        Y1(I)=Y(I)                                               215
        Y2(I)=Y(I)                                               216
10    CONTINUE                                                   217
      CALL RUNGE(N,H  ,T    ,Y1,DGL)                             218
      CALL RUNGE(N,H/2,T    ,Y2,DGL)                             219
      CALL RUNGE(N,H/2,T+H/2,Y2,DGL)                             220
      EPS=0.D0                                                   221
      DO 20 I=1,N                                                222
        RR= 16*ABS(Y2(I)-Y1(I))/15                               223
        IF (RR.GT.EPS) EPS=RR                                    224
20    CONTINUE                                                   225
      RELA=MAX( (EPS/TOL)**.2D0 , 1.D-8)                         226
      HMAX=H/RELA                                                227
      IF (ABS(HMAX).LT.TOL) THEN                                 228
        IERR=1                                                   229
        GOTO 2                                                   230
      END IF                                                     231
      IF (RELA.GT.2) THEN                                        232
        H=HMAX                                                   233
        TNEW=T+H                                                 234
        GOTO 1                                                   235
      END IF                                                     236
2     DO 30 I=1,N                                                237
        Y(I)=Y2(I)                                               238
30    CONTINUE                                                   239
      IF (RELA.LT.1) H=2*H                                       240
      T=TNEW                                                     241
      TNEW=T+H                                                   242
      IF (ABS(TNEW-T0).GE.ABS(T1-T0)) THEN                       243
        TNEW=T1                                                  244
        H=TNEW-T                                                 245
      END IF                                                     246
      IF (T.NE.T1) GOTO 1                                        247

      END                                                        248
```

Fig. 7.5. Subroutine FINTEG

choose the units for distance and time so that the y_i do not differ by too many powers of ten. For the unit of distance we shall use 1000 km and for the unit of time one hour; then for the physically most interesting flight paths the y_i differ by only a few orders of magnitude.

Like the subroutine RUNGE from Chap. 5, FINTEG also needs as parameter the name of another subroutine which computes the right-hand sides of the system of differential equations. We again call this parameter DGL (see line 200 in Fig. 7.5).

To integrate the equations of motion for the mesh widths h and $h/2$, FINTEG calls the subroutine RUNGE (lines 218–220). In the following DO-loop ε_h is computed (lines 222–225). To ensure that in the calculation of h_{max} division by zero is not attempted, the quotient h/h_{max} is set equal to at least 10^{-8} (lines 226–227). In line 232 the choice of cases mentioned in Sect. 7.2.1 is introduced. If the absolute value of $h/2$ is greater than h_{max}, h is set equal to $2h_{max}$ and the integration is tried again with the new mesh width (lines 233–236). Otherwise the $y_{i,h/2}(t+h)$ are taken as the new values of the y_i (lines 237–239). If h_{max} was greater than h, then h is doubled in the next integration step (line 240). In the lines 241 to 246 h and $t+h$ are computed for the next integration step. Here one has to test whether $t+h$ would overshoot t_1. Since t_1 may be smaller than t_0, this can be programmed only in a rather roundabout way (line 243). If the integration has already come to the end of the interval, FINTEG is concluded; otherwise one returns to Label 1 (line 247).

In order to prevent FINTEG computing for ever, without reaching the goal, we have to specify a lower limit for $|h|$. Otherwise it may happen that $t+h$ and t become numerically equal and the integration is no longer getting anywhere. In our example we can use ε_{max} as lower limit for $|h|$. The variable IERR reports back to the calling program whether h_{max} has become too small. At the beginning it holds the value zero (line 208). If $|h_{max}|$ becomes less than ε_{max}, then h is no longer made smaller, and IERR takes the value unity (lines 228–231).

7.3.3 Coordinate Transformation

The transformation of the position coordinates is carried out in subroutine TFX. Fig. 7.6 shows the FORTRAN notation for the quantities used in Sect. 7.2.2. The program TFX is listed in Fig. 7.7. The variable NDIR indicates the direction of the desired transformation. If NDIR is greater than 0, it is a matter of a trans-

Physical title	FORTRAN title	Physical title	FORTRAN title
r_d	RD	t	T
φ_d	PHID	ω	OMEGA
x_d	XD	ω_r	OMEGAR
y_d	YD	r_c	RCENT
r	Z1	ωt	PHIC
φ	Z2	x_c	XC
x	X	y_c	YC
y	Y		

Fig. 7.6. Notation employed in subroutine TFX

```
          SUBROUTINE TFX(NDIR,T,RD,PHID,Z1,Z2,XD,YD)              300

       IMPLICIT DOUBLE PRECISION(A-H,O-Z)                         301
       COMMON/SYST/G1,R1,G2,R2,OMEGA,OMEGAR,RCENT,PLOTR,NCOR,PI   302
*
       PHIR=OMEGAR*T                                              303
*      Computation of the centre in Cartesian coordinates
       PHIC=OMEGA*T                                               304
       CALL CARTES(RCENT,PHIC,XC,YC)                              305
       IF (NDIR.GT.0) THEN                                        306
*      Transformation into the canonical coordinate system
         CALL CARTES(RD,PHID,XD,YD)                               307
         CALL ROTAT(XD,YD,PHIR,X2,Y2)                             308
         CALL TRANSL(X2,Y2,XC,YC,X,Y)                             309
         CALL POLAR (X,Y,Z1,Z2)                                   310
       ELSE                                                       311
*      Transformation into the input/output system
         CALL CARTES(Z1,Z2,X,Y)                                   312
         CALL TRANSL(X,Y,-XC,-YC,X2,Y2)                           313
         CALL ROTAT(X2,Y2,-PHIR,XD,YD)                            314
         CALL POLAR(XD,YD,RD,PHID)                                315
       END IF                                                     316
       END                                                        317
```

Fig. 7.7. Subroutine TFX

formation into the coordinate system 0, otherwise a transformation into one of the input/output coordinate systems. For the sake of clarity the conversion between Cartesian and polar coordinates, the displacement and the rotation are carried out in separate subroutines.

In preparation for the transformation the displacement vector r_c is computed in Cartesian coordinates (lines 304 and 305). If the transformation required is from one of the coordinate systems for input/output into coordinate system 0, then the space craft's position r_d is converted into Cartesian coordinates (line 307), transformed according to (7.24) (lines 308 and 309), and the result converted back into polar coordinates (line 310). In a transformation from coordinate system 0 into one of the coordinate systems for input/output, all the transformation steps are made in the reverse direction and in the reverse order (lines 312–315).

The transformation of the velocity coordinates (from 7.27) is carried out in subroutine TFV. It assumes that the positional coordinates r, φ, x_d and y_d have already been computed in TFX. Besides these quantities known from TFX it also uses others, which are listed in Fig. 7.8.

In preparation for the transformation (7.27) one first computes dr_c/dt and v_r from (7.28) and (7.29b) (lines 404–409 in Fig. 7.9). The transformation from the coordinate systems for the input/output into the coordinate system 0 is carried out in lines 412 to 415. In the two following lines the canonical momentum coordinates are computed from (7.30). The transformation from the coordinate system 0 into one of the coordinate systems for the input/output is programmed in lines 419 to 424.

Physical title	FORTRAN title	Physical title	FORTRAN title
v_d	VD	$v_{r,x}$	VXR
ψ_d	PSID	$v_{r,y}$	VYR
p_r	Z3	dx_c/dt	VXC
p_φ	Z4	dy_c/dt	VYC

Fig. 7.8. Additional notation in subroutine TFV

```
      SUBROUTINE TFV(NDIR,T,VD,PSID,Z1,Z2,Z3,Z4,XD,YD)          400
      IMPLICIT DOUBLE PRECISION(A-H,O-Z)                         401
      COMMON/SYST/G1,R1,G2,R2,OMEGA,OMEGAR,RCENT,PLOTR,NCOR,PI   402
*
      PHIR=OMEGAR*T                                              403
* Computation of the rotational velocity in Cartesian coordinates
      VXR=-OMEGAR*YD                                             404
      VYR= OMEGAR*XD                                             405
* Computation of the velocity of the centre in Cartesian coordinates
      PHIC=OMEGA*T                                               406
      CALL CARTES(RCENT,PHIC,DX,DY)                              407
      VXC=-OMEGA*DY                                              408
      VYC= OMEGA*DX                                              409
* To convert canonical position coordinates into Cartesian
      CALL CARTES(Z1,Z2,X,Y)                                     410
      IF (NDIR.GT.0) THEN                                        411
* Transformation into the canonical coordinate system
        CALL CARTES(VD,PSID,VX1,VY1)                             412
        CALL TRANSL(VX1,VY1, VXR , VYR ,VX2,VY2)                 413
        CALL ROTAT (VX2,VY2,  PHIR     ,VX3,VY3)                 414
        CALL TRANSL(VX3,VY3, VXC , VYC ,VX4,VY4)                 415
        Z3=(VX4*X+VY4*Y)/Z1                                      416
        Z4=(VY4*X-VX4*Y)                                         417
      ELSE                                                       418
* Transformation into the input/output system
        VX4=(Z1*Z3*X-Z4*Y)/(Z1*Z1)                               419
        VY4=(Z1*Z3*Y+Z4*X)/(Z1*Z1)                               420
        CALL TRANSL(VX4,VY4,-VXC ,-VYC ,VX3,VY3)                 421
        CALL ROTAT (VX3,VY3, -PHIR     ,VX2,VY2)                 422
        CALL TRANSL(VX2,VY2,-VXR ,-VYR ,VX1,VY1)                 423
        CALL POLAR (VX1,VY1,VD,PSID)                             424
      END IF                                                     425
      END                                                        426
```

Fig. 7.9. Subroutine TFV

For completeness the subroutines ROTAT, TRANSL, CARTES and POLAR are collected together in Fig. 7.10. Since these are very simple we dispense with a description.

7.3.4 Main Program

The main program controls the calling of FINTEG, the input of the initial conditions and the data of the earth-moon system as well as the graphical output of the computed positions of the space craft.

```
      SUBROUTINE ROTAT(X,Y,PHI,XOUT,YOUT)                  500
      IMPLICIT DOUBLE PRECISION(A-H,O-Z)                   501
      COPH=COS(PHI)                                        502
      SIPH=SIN(PHI)                                        503
      XOUT=X*COPH-Y*SIPH                                   504
      YOUT=Y*COPH+X*SIPH                                   505
      END                                                  506

      SUBROUTINE TRANSL(X,Y,DX,DY,XOUT,YOUT)               507
      IMPLICIT DOUBLE PRECISION(A-H,O-Z)                   508
      XOUT=X+DX                                            509
      YOUT=Y+DY                                            510
      END                                                  511

      SUBROUTINE CARTES(R,PHI,X,Y)                         512
      IMPLICIT DOUBLE PRECISION(A-H,O-Z)                   513
      X=R*COS(PHI)                                         514
      Y=R*SIN(PHI)                                         515
      END                                                  516

      SUBROUTINE POLAR(X,Y,R,PHI)                          517
      IMPLICIT DOUBLE PRECISION(A-H,O-Z)                   518
      R=SQRT(X*X+Y*Y)                                      519
      IF (R.GT.0) THEN                                     520
        PHI=ATAN2(Y,X)                                     521
      ELSE                                                 522
        PHI=0                                              523
      END IF                                               524
      END                                                  525
```

Fig. 7.10. Subroutines ROTAT, TRANSL, CARTES and POLAR

Physical title	FORTRAN title	Physical title	FORTRAN title
π	PI	r_d	RD
(1 month in hours)	AMONTH	φ_d	PHID
ω	OMEGA	x_d	XD
$r_1 + r_2$	ABSTD	y_d	YD
m_2/m_1	GX	$d|r_d|/dt$	VD
r_1	R1	ψ_d	PSID
r_2	R2	$y_1 = r$	Y(1)
γm_1	G1	$y_2 = \varphi$	Y(2)
γm_2	G2	$y_3 = p_r$	Y(3)
t, t_0	T	$y_4 = p_\varphi$	Y(4)
t_1	TE	ε_{max}	TOL
		$t_1 - t_0$	DELTAT

Fig. 7.11. Notation employed in main program KAP7

```
      PROGRAM KAP7                                             600
      IMPLICIT DOUBLE PRECISION(A-H,O-Z)                       601
      PARAMETER (AMONTH=648.D0,DIST=384.D0)                    602
      PARAMETER (TOLMAX=1.D-3)                                 603
      DIMENSION Y(4)                                           604
      COMMON/SYST/G1,R1,G2,R2,OMEGA,OMEGAR,RCENT,PLOTR,NCOR,PI 605
      EXTERNAL DGL7                                            606
*     (Variables for input/output)

      PI=4*ATAN(1.D0)                                          607
      OMEGA=2*PI/AMONTH                                        608

*     (Input of TOL, GX, PLOTR, choice of coordinate system
*      and determination of OMEGAR, RCENT)
      R1=DIST*GX/(1+GX)                                        609
      R2=DIST/(1+GX)                                           610
      G1=R2*OMEGA**2*DIST**2                                   611
      G2=GX*G1                                                 612
      T=0                                                      613

*     (Input of RD,PHID,VD,PSIG,DELTAT)

      CALL TFX(1,T,RD,PHID,Y(1),Y(2),XD,YD)                    614
      CALL TFV(1,T,RD,PHID,VD,PSID,Y(1),Y(2),Y(3),Y(4),XD,YD)  615
      NIT=0                                                    616

*     (Graphical output of the initial positions of earth, moon,
*      space craft)

4     CONTINUE                                                 617
      IF (NIT.LT.24) THEN                                      618
        NIT=NIT+1                                              619
        TE=T+DELTAT                                            620
        CALL FINTEG(T,TE,4,Y,TOL,DGL7,IERR)                    621
        T=TE                                                   622
        CALL TFX(-1,T,RD,PHID,Y(1),Y(2),XD,YD)                 623

*     (Graphical output of the new positions)

        GOTO 4                                                 624
      END IF                                                   625

*     (Control of remainder of flight, e.g. change of coordinate system,
*      change of velocity, further flights, new start)

      END                                                      626
```

Fig. 7.12. Numerical portion of the main program KAP7

Figure 7.12 presents the numerical portion of the main program. In lines 607 and 608 are specified two quantities needed again and again in various subroutines, and these are therefore stored in the COMMON block /SYST/. The computation of the number π from the arctan (line 607) has the advantage that it is less susceptible to typing errors in the program than explicitly writing out PI = 3.1415926 ..., and that it also uses the full precision of the DOUBLE PRECISION quantities.

75

After the input of the mass ratio m_2/m_1, r_1 and r_2 are computed from (7.8) in lines 609 and 610. Since the earth and the moon are moving round each other under the influence of gravitation, there is also a connection between the angular velocity ω, the earth mass m_1 and the moon mass m_2. For these bodies to stay on their circular orbits, the gravitational force and the centrifugal force must be equal:

$$\frac{\gamma m_1 m_2}{(r_1+r_2)^2} = m_1\omega^2 r_1 = m_2\omega^2 r_2 \quad , \tag{7.37a}$$

from which it follows that

$$\gamma m_1 = \omega^2 r_2 (r_1+r_2)^2 \quad . \tag{7.37b}$$

Since we wish to specify in our program the earth-moon distance and the orbital period of the moon around the earth, we must adjust the earth mass m_1 whenever we change the mass ratio m_2/m_1. We compute γm_1 according to (7.37b) in line 611 and the resulting value for γm_2 in line 612.

After the input of the initial values for position and velocity of the space craft these must be converted into canonical coordinates (lines 614 and 615). The integration of the equations of motion follows. After the computation of each new flight path point (line 621) the new positions of the earth, moon and space craft are plotted. This is repeated 24 times (lines 616 to 619 and 624). Thereafter one has the possibility of excercising control over the course of events, e.g. by changing the velocity of the space craft or by choosing a different system of coordinates. If one has, for example, specified a time step of one hour one can make a new choice after each day of the flight.

The alphanumeric input generally retains the scheme of the previous chapters. The choice of coordinate system can be made at two different places in the main program, and is therefore carried out in a separate subroutine called GETCOR. In GETCOR values are also assigned to the variables RCENT and OMEGAR from the COMMON block /SYST/, which are needed in the subroutines TFX and TFV. The corresponding portion of the program is shown in Fig. 7.13. Here NCOR indicates the number of the coordinate system according to Sect. 7.1. The remaining notation corresponds to that of the subroutines TFX and TFV.

```
      IF (NCOR.LE.2) THEN                              700
         OMEGAR=0                                      701
      ELSE                                             702
         OMEGAR=OMEGA                                  703
      END IF                                           704
      IF (NCOR.EQ.0.OR.NCOR.EQ.3) THEN                 705
         RCENT=0                                       706
      ELSE IF (NCOR.EQ.1.OR.NCOR.EQ.4) THEN            707
         RCENT=-R1                                     708
      ELSE                                             709
         RCENT= R2                                     710
      END IF                                           711
```

Fig. 7.13. Assignment of values to RCENT and OMEGAR

In order to provide the possibility of exerting control during the graphical output, one needs the subroutine GINKEY from the graphics package, which has not so far been mentioned. It makes possible the input of an individual character while the graphics is active. The character does not appear on the screen, so that the graphics is not changed by the input.

The call is

CALL GINKEY(NOTE, CH)

GINKEY first displays the text of NOTE on the top margin of the screen. One should choose a text that tells the user what he is to do. The parameter CH must be a CHARACTER variable. It contains the specified character from the call. Function and cursor keys are denoted by a code of two letters. The codes are to be found, for example, in the Appendix of IBM's BASIC Manual. If one wishes to use such keys, CH should be of type CHARACTER*2.

The subroutine GINIT is used in order to switch between graphical and alphanumerical output. The call is:

CALL GINIT(1)   (to open the graphics)

CALL GINIT(−1)   (to close the graphics) .

The difference from GOPEN and GCLOSE is that, apart from opening and closing, nothing happens. In the opening the frame, axes and labels are not drawn. In the closing, no text is output and no key depression is expected. The use of GINIT is appropriate, for example, when the graphics is to be ended immediately in response to a key depression demanded by GINKEY.

In order to output the position of the space craft, the subroutine GMARK is used. The call is

CALL GMARK(X, Y, ICOLOR, ISTYLE)

The DOUBLE PRECISION quantities X and Y indicate the position at which a marker is to be shown. As in GMOVE and GDRAW, ICOLOR indicates the required colour. The precise form of the marker is indicated by ISTYLE, according to the code:

ISTYLE=0:   small dot
ISTYLE=1:   large dot
ISTYLE=2:   plus sign
ISTYLE=3:   asterisk
ISTYLE=4:   square
ISTYLE=5:   cross
ISTYLE=6:   diamond

The positions of the earth and the moon are indicated by means of the subroutine GCIRCL. The call is

CALL GCIRCL(X, Y, R, ICOLOR, ISTYLE)

GCIRCL draws a circle with radius R and centre (X,Y). Here X, Y and R are DOUBLE PRECISION quantities. The radius R is in the same units as the abscissa. The required colour is again specified by ICOLOR. The value of ISTYLE indicates whether the circle is to be hollow, filled or hatched, according to the code:

ISTYLE=1: hollow
ISTYLE=2: filled
ISTYLE=3: densely hatched
ISTYLE=4: moderately hatched
ISTYLE=5: thinly hatched
ISTYLE=6: densely cross-hatched
ISTYLE=7: moderately cross-hatched
ISTYLE=8: thinly cross-hatched

The subroutine GWRITE outputs a text on the graphics screen. The call is:

CALL GWRITE(1,XNOTE,YNOTE,NOTE,ICOLOR)

The text to be output must be specified in NOTE as a CHARACTER*(*) expression. ICOLOR states the required colour. The x-coordinate of the text is given by the DOUBLE PRECISION quantity XNOTE, the y-coordinate by YNOTE. Unlike the previous subroutines, the coordinates here indicate the position, not in the physical units, but as a percentage of the screen dimensions. The bottom left-hand corner of the screen has the coordinates (0,0), the top right-hand corner the coordinates (100,100). If one sets the first parameter of GWRITE at a value other than unity, one can also use other coordinate systems. A complete description of the facility is to be found on the diskette.

We use GWRITE in order to output the time t and the distance r_d of the space craft from the origin of coordinates. Since t and r_d are DOUBLE PRECISION quantities, but for the parameter NOTE we need a CHARACTER expression, we have to make a conversion. This is achieved by using variables of type CHARACTER as the argument of a WRITE statement. By

CHARACTER*8 GCHAR

...

WRITE(GCHAR,'(F8.2)') T

the CHARACTER variable GCHAR takes the value of T, expressed as a floating point number with two places after the decimal point.

The complete program is stored on the diskette in the file KAP-07.FOR.

7.4 Exercises

7.4.1 Test the accuracy of the integration by following the motion of a space craft in a close terrestrial orbit for a few days, and then integrating backwards. How small must TOL be to regain the specified starting position at time 0?

7.4.2 Space craft in terrestrial orbits must obey Kepler's Laws. Can you test this with the aid of the graphical representation of the space craft's path?

7.4.3 Try to find the starting conditions for hitting the moon. Make the moon somewhat heavier than in reality, in order to make hitting it easier. It is advantageous to choose the initial conditions as if one were starting from a circular orbit round the earth.

7.4.4 Try to find a path on which the lunar gravitation swings the space craft round so that it returns to the immediate vicinity of the earth. How sensitive is such a path to slight variations of the initial conditions? Try to bring the space craft in the immediate vicinity of the moon into a lunar orbit. By how much must the velocity be decreased to achieve this? Estimate what total changes of velocity must be produced by the jets in an Apollo moon mission with flight to the moon, entry into lunar orbit, landing on the moon and return to earth.

7.4.5 How large are the maximal orbital radii for which orbits round the earth and the moon are stable?

7.5 Solutions to the Exercises

7.5.1 If one calculates the path of a space craft in a terrestrial orbit for a duration of about 3 days, one needs an error limit TOL of 0.00001 or less, in order to return by a reversed motion back to the initial conditions for position and velocity. With TOL = 0.00001 one will still have an error of up to 1 degree for φ and ψ. Only with an error limit of about 0.0000001 will the reciprocal path be almost identical with the original one.

7.5.2 According to Kepler's first law, the path of the space craft must be an ellipse, with the centre of the earth as one of the foci.

Kepler's second law states that the line from the centre of the earth to the space craft sweeps out equal areas in equal times. Since the position of the space craft is given at equal time intervals, Kepler's second law implies that the markers showing the positions of the space craft are spaced much more densely in the region of the points most distant from the earth than for the nearest points.

Kepler's third law states that the square of the major semi-axis is proportional to the cube of the orbital period. One can test this by comparing the orbital periods for circular and elliptical orbits with different major semi-axes.

7.5.3 and **7.5.4** A trajectory which passes very close to the moon and returns close to the earth is obtained for

$$m_2/m_1 = 0.02$$

and initial conditions (coordinate system 1):

$r_d = 6800 \, \text{km}$,
$\varphi_d = 235°$,
$v_d = 38\,882 \, \text{km/h}$,
$\psi_d = 325°$.

By slight modifications of these values one obtains trajectories which strike the moon or the earth.

If one changes the coordinate system in the vicinity of the moon, one sees that the space craft flies by the moon at about 9000 km distance with a velocity of about 5500 km/h.

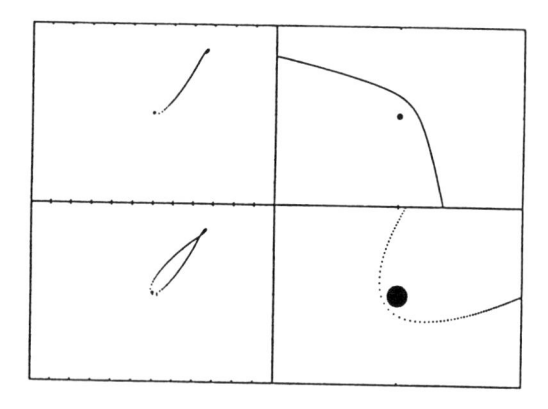

Fig. 7.14. Path of a space craft with reversal by the moon and return to the earth

In order to enter a lunar orbit one must reduce the velocity by at least 1500 km/h. If one reduces the velocity by about 3500 km/h one enters an orbit whose nearest point to the moon is in the immediate proximity of the lunar surface. Then the space craft flies by the lunar surface with a relative velocity of about 9500 km/h.

The execution of a moonflight with lunar landing and return to earth requires, in our model, the following velocity changes:

39000 km/h	investment in setting course for the moon
3500 km/h	retardation for lunar orbit
9500 km/h	landing on the moon
9500 km/h	take-off from the lunar surface
3500 km/h	acceleration for return journy to the earth

65000 km/h

By more skilful choice of the flight parameters, together with the use of the realistic moon mass, one can still save a little. However, the total energy requirement for the Apollo space craft was actually of the order of magnitude indicated. Remark: The retardation of the space craft on return to earth is accomplished by the terrestrial atmosphere.

7.5.5 With a realistic ratio of the moon mass to the earth mass (GX=0.012) the perturbation of terrestrial satellites becomes really significant at an orbital height of about 250 000 km. Circular orbits around the moon are already strongly perturbed at heights of about 20 000 km. In the real earth-moon system the perturbation of lunar orbits is much stronger still, because the mass distribution in the moon is not quite spherically symmetric, because of the eccentricity of the moon's orbit, and because of perturbations due to the sun. It is generally rather difficult to bring a lunar satellite into a stable orbit.

8. The Celestial Mechanics Three-body Problem

8.1 Formulation of the Problem

The flight of a space craft to the moon, such as we have dealt with in Chap. 7, is a special case of the three-body problem of celestial mechanics. The latter is understood to be the problem of calculating the motion of three point-like masses which move under the sole influence of the mutual gravitational forces.

We have treated the three-body problem of celestial mechanics with three limitations:

1. The third mass is so small that it does not influence the motion of the other two bodies.
2. The two large masses move in a circular orbit.
3. All three bodies move in a non-rotating plane.

When the first two conditions are fulfilled, one speaks of the "restricted three-body problem in celestial mechanics". We shall concern ourselves further with this problem, and shall also retain the limitation to motion in a non-rotating plane. Since we shall discuss a few examples which have nothing to do with the earth-moon system, we shall from now on drop the titles "earth" and "moon" for the masses m_1 and m_2.

In Chap. 7 we have carried out the calculation in an inertial system with a time-dependent interaction. For the input/output, however, we have already introduced coordinate systems which rotate with the two large bodies. We shall now use such a coordinate system for the calculation also. We choose the coordinate system 3 from the previous chapter. The origin of coordinates is the total centre of mass. The potential V is no longer explicitly time-dependent. The equations of motion therefore become more complicated by the appearance of new centrifugal and Coriolis terms. The Hamilton formalism does not assume that an inertial system is used in the definition of the position coordinates $q_k(t)$: it can also be used in a rotating system of coordinates.

In order to obtain Hamilton's equations in the rotating coordinate system, we first construct the Lagrange function. We carry over the notation from Chap. 7. In the calculation of the kinetic energy T for the mass μ we must take account of the additional motion resulting from the rotation of the coordinate system. We then have:

$$T = \frac{\mu}{2}\left(\left(\frac{dr}{dt}\right)^2 + r^2\left(\frac{d\varphi}{dt} + \omega\right)^2\right) \ . \tag{8.1}$$

The expression for the potential energy V of the mass μ depends only on the distances $s_i = |\boldsymbol{r}_i - \boldsymbol{r}|$ and is not influenced by the rotation of the coordinate system.

We obtain for the Lagrange function and for the canonical momenta p_k:

$$L = T - V = \frac{\mu}{2}\left(\left(\frac{dr}{dt}\right)^2 + r^2\left(\frac{d\varphi}{dt} + \omega\right)^2\right) + \frac{\gamma m_1 \mu}{s_1} + \frac{\gamma m_2 \mu}{s_2} \quad , \tag{8.2}$$

$$p_r = \mu \frac{dr}{dt} \quad , \tag{8.3a}$$

$$p_\varphi = \mu r^2 \left(\frac{d\varphi}{dt} + \omega\right) \quad . \tag{8.3b}$$

We can now determine the Hamilton function from (7.3):

$$H' = \mu \frac{dr}{dt}\frac{dr}{dt} + \mu r^2 \left(\frac{d\varphi}{dt} + \omega\right)\frac{d\varphi}{dt} - L$$

$$= \frac{\mu}{2}\left(\left(\frac{dr}{dt}\right)^2 + r^2\left(\frac{d\varphi}{dt}\right)^2 - r^2\omega^2\right) - \frac{\gamma m_1 \mu}{s_1} - \frac{\gamma m_2 \mu}{s_2} \quad , \tag{8.4}$$

$$H = \frac{p_r^2}{2\mu} + \frac{p_\varphi^2}{2\mu r^2} - p_\varphi \omega - \frac{\gamma m_1 \mu}{s_1} - \frac{\gamma m_2 \mu}{s_2} \quad . \tag{8.5}$$

H differs from the Hamilton function (7.15) by the term $-p_\varphi \omega$. This affects only the second of the four equations of motion (7.16), the others remaining unchanged:

$$\frac{dr}{dt} = \frac{p_r}{\mu} \quad , \tag{8.6a}$$

$$\frac{d\varphi}{dt} = \frac{p_\varphi}{\mu r^2} - \omega \quad , \tag{8.6b}$$

$$\frac{dp_r}{dt} = \frac{p_\varphi^2}{\mu r^3} - \frac{\gamma m_1 \mu}{s_1^3}(r - r_1 \cos(\varphi - \varphi_1)) - \frac{\gamma m_2 \mu}{s_2^3}(r - r_2 \cos(\varphi - \varphi_2)) \quad , \tag{8.6c}$$

$$\frac{dp_\varphi}{dt} = -\frac{\gamma m_1 \mu}{s_1^3} r r_1 \sin(\varphi - \varphi_1) - \frac{\gamma m_2 \mu}{s_2^3} r r_2 \sin(\varphi - \varphi_2) \quad . \tag{8.6d}$$

Since our coordinate system rotates with m_1 and m_2, the angles φ_1 and φ_2 are no longer time-dependent. We choose the spatial orientation so that m_1 and m_2 lie on the x-axis:

$$\varphi_1 = \pi \quad , \quad \varphi_2 = 0 \quad . \tag{8.7}$$

Already in the 17th century mathematicians and astronomers were interested in steady solutions of the restricted three-body problem of celestial mechanics, i.e. in solutions which fulfil the conditions

$$\frac{dr}{dt} \equiv 0 \quad , \quad \frac{d\varphi}{dt} \equiv 0 \quad ,$$

$$\frac{dp_r}{dt} \equiv 0 \quad , \quad \frac{dp_\varphi}{dt} \equiv 0 \quad . \tag{8.8}$$

Since we have already derived the equations of motion (8.6), we shall start out from these in order to find solutions which fulfil (8.8). We again set μ equal to unity. The first conditions can easily be evaluated:

$$\frac{dr}{dt} = p_r = 0 \quad , \tag{8.9a}$$

$$\frac{d\varphi}{dt} = p_\varphi - \omega r^2 = 0 \quad , \quad \text{and hence} \quad p_\varphi = \omega r^2 \quad . \tag{8.9b}$$

For the other two conditions we use the relation (7.37) between m_1, m_2 and ω derived in Chap. 7:

$$\gamma m_1 = \omega^2 r_2 (r_1 + r_2)^2 \quad , \tag{8.10a}$$

and correspondingly

$$\gamma m_2 = \omega^2 r_1 (r_1 + r_2)^2 \quad . \tag{8.10b}$$

Taking account of (8.6d) and (8.7) we can then write the fourth of the conditions (8.8) as

$$\frac{dp_\varphi}{dt} = r_1 r_2 \omega^2 (r_1 + r_2)^2 r \sin\varphi \left(\frac{1}{s_1^3} - \frac{1}{s_2^3} \right) \equiv 0 \quad . \tag{8.11}$$

Ignoring the trivial solution $r = 0$, for which (8.6c) cannot be evaluated, (8.11) can be fulfilled if

$$\sin\varphi = 0 \quad , \tag{8.12a}$$

or if

$$s_1 = s_2 \quad . \tag{8.12b}$$

In the first case the three bodies lie in a line, in the second case they form a triangle. If in the first case one wishes also to evaluate the third of the conditions (8.8), one must substitute (8.12a) and (8.9b) in (8.6c). One obtains an algebraic equation of the fifth degree for r, which has three real solutions. From an astronomical standpoint, however, these are not particularly interesting, since they involve unstable solutions.

The second case (8.12b) is more interesting. By using (8.10b) and (8.7) one can rewrite the equation of motion (8.6c) as

$$\frac{dp_r}{dt} = \frac{p_\varphi^2}{r^3} - (r_1 + r_2)^2 \omega^2 \left(\frac{r_2}{s_1^3}(r + r_1 \cos\varphi) + \frac{r_1}{s_2^3}(r - r_2 \cos\varphi) \right). \tag{8.13}$$

In order to evaluate the third condition from (8.8), we have to substitute the relations (8.9b) and (8.12b) into (8.13). Since according to (8.12b) s_1 and s_2 are equal, the terms involving $\cos\varphi$ in (8.13) vanish. In what follows we shall denote s_1 and s_2 by s. Accordingly:

$$\frac{dp_r}{dt} = \omega^2 r - \frac{(r_1+r_2)^3 \omega^2 r}{s^3} = 0 \tag{8.14a}$$

and therefore

$$s = r_1 + r_2 \quad . \tag{8.14b}$$

Since $r_1 + r_2$ is the distance between the two large masses, (8.14b) means that the three masses form an equilateral triangle. For the orientation of the triangle relative to the axis of rotation there are two possibilities (see Fig. 8.1). The triangle solutions are stable if m_2 is significantly smaller than m_1. For nearly equal masses the stability is not ensured in every case.

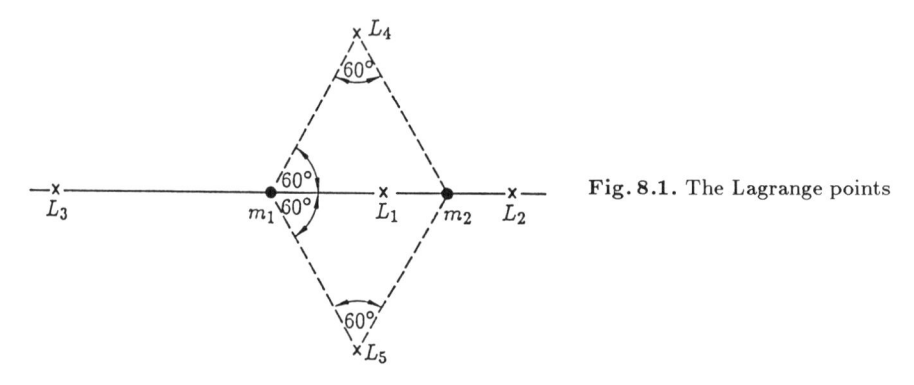

Fig. 8.1. The Lagrange points

The five positions of the mass μ for which steady solutions of the three-body problem exist are named Lagrange points, and labelled L_1 to L_5, after the famous French mathematician who first discovered them. Stable solutions are found at L_4 and L_5, see Fig. 8.1.

Since the beginning of this century a group of asteroids, called "Trojan" asteroids, has been known which move in the path of the planet Jupiter and which stay about 60 degrees ahead of, or behind, it. They are located in the immediate vicinity of the equilibrium points L_4 and L_5 and execute kidney-shaped oscillations about these points, which are known as libration oscillations. In the system of the moons of Saturn also "Trojan" satellites are observed, which follow the same path as the larger moon, and about 60 degrees ahead of, or behind, it.

A further interesting solution of the three-body problem is the so-called "horseshoe" orbit. There a small moon executes a horseshoe oscillation about the path of a larger moon, including the Lagrange points L_3, L_4 and L_5. A variant of the horseshoe orbit is said to have been observed in the system of Saturn's moons.

The equilibrium solutions of the three-body problem of celestial mechanics are only one example of the many resonance phenomena which play a role in the celestial mechanics of the solar system. Another well-known example are the resonant perturbations in the asteroid belt. Between the orbits of Mars and Jupiter there are several thousand asteroids with sizes between one and a few hundred kilometres. In the distribution of the asteroids there are conspicuous gaps. Asteroids whose orbital periods bear a simple whole-number ratio to that of Jupiter, e.g. $1:3$ or $2:5$, are almost totally absent. The origin of the gap for $1:3$ is explained with some certainty. With this orbital period the asteroids find themselves repeatedly at the same position relative to Jupiter, so that the perturbations of the orbits of the asteroids by Jupiter can build up. The eccentricity of the asteroidal orbits increases and after some time becomes so great that the asteroids cross the orbit of Mars, and by a close encounter with Mars their orbits are completely changed [8.1]. Resonances can accordingly make an orbit unstable, but they can also stabilise it, as demonstrated by the Trojan asteroids.

8.2 Mathematical Method

For the solution of Hamilton's equations (8.6) we apply the Runge-Kutta method discussed in Chap. 7 with mesh width adjustment.

8.3 Programming

The programming is in general identical with that of Chap. 7. We therefore discuss only the differences from Chap. 7 which arise.

Although our study is no longer directed at the earth-moon system, we shall still assume that the masses m_1 and m_2 are 384 000 km apart and circle each other in 648 hours. This gives us a simple method of testing whether we have made mistakes in writing the program, since the results should not be different now that Hamilton's equations are solved in coordinate system 3, instead of coordinate system 0. By a new definition of kilometre and hour we can transfer our results to other examples of the restricted three-body problem of celestial mechanics.

For the solution of the equations of motion (8.6) we again number the coordinates as in (7.33). Then with $\mu = 1$ we have

$$\frac{dy_i}{dt} = f_i(y_1, y_2, y_3, y_4; t) \quad , \quad i = 1, \ldots, 4 \quad , \tag{8.15a}$$

$$y_1 = r \quad , \quad y_2 = \varphi \quad , \tag{8.15b}$$

$$y_3 = p_r \quad , \quad y_4 = p_\varphi \quad ,$$

$$f_1 = y_3 \quad , \tag{8.16a}$$

$$f_2 = \frac{y_4}{y_1^2} - \omega \quad , \tag{8.16b}$$

$$f_3 = \frac{y_4^2}{y_1^3} - \frac{\gamma m_1}{s_1^3}(y_1 - r_1 \cos(y_2 - \varphi_1))$$
$$- \frac{\gamma m_2}{s_2^3}(y_1 - r_2 \cos(y_2 - \varphi_2)) \quad , \tag{8.16c}$$

$$f_4 = -\frac{\gamma m_1}{s_1^3} y_1 r_1 \sin(y_2 - \varphi_1) - \frac{\gamma m_2}{s_2^3} y_1 r_2 \sin(y_2 - \varphi_2) \quad . \tag{8.16d}$$

We compute the function values f_i in the subroutine DGL8 (see Fig. 8.2). It differs from DGL7 in line 105, where the additional term from (8.16b) is taken into account, and in lines 107 and 108, since the angle φ_2 vanishes according to (8.7). In order to facilitate the comparison the same numbering of the lines has been used in Fig. 8.2 as in Fig. 7.3.

```
      SUBROUTINE DGL8(T,Y,F)                                        100

      IMPLICIT DOUBLE PRECISION(A-H,O-Z)                            101
      DIMENSION F(4),Y(4)                                           102
      COMMON/SYST/G1,R1,G2,R2,OMEGA,OMEGAR,RCENT,PLOTR,NCOR,PI      103

      F(1)=Y(3)                                                     104
      F(2)=Y(4)/Y(1)**2-OMEGA                                       105
*     (line 106 omitted)
      COPH=COS(Y(2))                                                107
      SIPH=SIN(Y(2))                                                108
      S1=SQRT( R1*R1 + Y(1)*Y(1) + 2*R1*Y(1)*COPH )                 109
      S2=SQRT( R2*R2 + Y(1)*Y(1) - 2*R2*Y(1)*COPH )                 110
      S13=S1**3                                                     111
      S23=S2**3                                                     112
      F(3)=(Y(4)**2)/(Y(1)**3)                                      113
     &     -G1*(Y(1)+R1*COPH)/S13                                   114
     &     -G2*(Y(1)-R2*COPH)/S23                                   115
      F(4)= G1*(Y(1)*R1*SIPH)/S13                                   116
     &     -G2*(Y(1)*R2*SIPH)/S23                                   117
      END                                                           118
```

Fig. 8.2. Subroutine DGL8

Since m_1 and m_2 are at rest in coordinate system 3, the displacement vector r_c in the coordinate transformation is no longer time-dependent. We can therefore simplify subroutine TFX (lines 304 and 305 of Fig. 8.3). In order to avoid confusion we call the new subroutine TFXR.

The alterations in subroutine TFV, which we now call TFVR (Fig. 8.4), are more important. Since r_c no longer varies with time, dr_c/dt is equal to zero, and a part of the transformation (previously lines 406–409, 415 and 421) can

```
        SUBROUTINE TFXR(NDIR,T,RD,PHID,Z1,Z2,XD,YD)                      300

        IMPLICIT DOUBLE PRECISION(A-H,O-Z)                                301
        COMMON/SYST/G1,R1,G2,R2,OMEGA,OMEGAR,RCENT,PLOTR,NCOR,PI          302

        PHIR=OMEGAR*T                                                     303
*       Computation of the centre in Cartesian coordinates
        XC=RCENT                                                          304
        YC=0                                                              305
        IF (NDIR.GT.0) THEN                                               306
*       Transformation into the canonical coordinate system
          CALL CARTES(RD,PHID,XD,YD)                                      307
          CALL ROTAT(XD,YD,PHIR,X2,Y2)                                    308
          CALL TRANSL(X2,Y2,XC,YC,X,Y)                                    309
          CALL POLAR (X,Y,Z1,Z2)                                          310
        ELSE                                                              311
*       Transformation into the input/output system
          CALL CARTES(Z1,Z2,X,Y)                                          312
          CALL TRANSL(X,Y,-XC,-YC,X2,Y2)                                  313
          CALL ROTAT(X2,Y2,-PHIR,XD,YD)                                   314
          CALL POLAR(XD,YD,RD,PHID)                                       315
        END IF                                                            316
        END                                                               317
```

Fig. 8.3. Subroutine TFXR

```
        SUBROUTINE TFVR(NDIR,T,VD,PSID,Z1,Z2,Z3,Z4,XD,YD)                 400

        IMPLICIT DOUBLE PRECISION(A-H,O-Z)                                401
        COMMON/SYST/G1,R1,G2,R2,OMEGA,OMEGAR,RCENT,PLOTR,NCOR,PI          402

        PHIR=OMEGAR*T                                                     403
*       Computation of the rotational velocity in Cartesian coordinates
        VXR=-OMEGAR*YD                                                    404
        VYR= OMEGAR*XD                                                    405
*       (lines 406-409 omitted)
*       To convert canonical position coordinates into Cartesian
        CALL CARTES(Z1,Z2,X,Y)                                            410
        IF (NDIR.GT.0) THEN                                               411
*       Transformation into the canonical coordinate system
          CALL CARTES(VD,PSID,VX1,VY1)                                    412
          CALL TRANSL(VX1,VY1, VXR , VYR ,VX2,VY2)                        413
          CALL ROTAT (VX2,VY2,    PHIR      ,VX4,VY4)                     414
*       (line 415 omitted)
          Z3=(VX4*X+VY4*Y)/Z1                                             416
          Z4=(VY4*X-VX4*Y)+OMEGA*Z1*Z1                                    417
        ELSE                                                              418
*       Transformation into the input/output system
          VX4=(Z1*Z3*X-Z4*Y)/(Z1*Z1)+OMEGA*Y                              419
          VY4=(Z1*Z3*Y+Z4*X)/(Z1*Z1)-OMEGA*X                              420
*       (line 421 omitted)
          CALL ROTAT (VX4,VY4,  -PHIR      ,VX2,VY2)                      422
          CALL TRANSL(VX2,VY2,-VXR  ,-VYR  ,VX1,VY1)                      423
          CALL POLAR (VX1,VY1,VD,PSID)                                    424
        END IF                                                            425
        END                                                               426
```

Fig. 8.4. Subroutine TFVR

be omitted. Two other lines have to be modified, because the displacement by dr_c/dt vanishes (lines 414 and 422).

In the computation of the canonical momenta we have to take into account the fact that according to (8.3b) the definition of p_φ has changed. Instead of (7.30) we now have:

$$p_r = \frac{dr}{dt} = \frac{d}{dt}\sqrt{x^2 + y^2} = \frac{1}{r}\left(x\frac{dx}{dt} + y\frac{dy}{dt}\right) , \qquad (8.17a)$$

$$p_\varphi = x\frac{dy}{dt} - y\frac{dx}{dt} + r^2\omega . \qquad (8.17b)$$

For the transformation from the coordinate system 3 into the input/output system the system of equations (8.17) has to be inverted. Instead of (7.31) we now have:

$$\frac{dx}{dt} = \frac{rp_r x - yp_\varphi}{r^2} + y\omega , \qquad (8.18a)$$

$$\frac{dy}{dt} = \frac{rp_r y + xp_\varphi}{r^2} - x\omega . \qquad (8.18b)$$

The corresponding alterations occur in the lines 417 to 420.

The alterations in the numerical part of the main program consist in replacing DGL7, TFX and TFV by DGL8, TFXR and TFVR

The subroutine GETCOR determines the angular velocity ω_r with which the coordinate system for input/output rotates, relative to the coordinate system 3. For the coordinate systems 0 to 2, ω_r is now equal to $-\omega$, whilst for the coordinate systems 3 to 5 it is equal to zero. The corresponding changes are displayed in Fig. 8.5.

```
      IF (NCOR.LE.2) THEN                                    700
         OMEGAR=-OMEGA                                       701
      ELSE                                                   702
         OMEGAR=0                                            703
      END IF                                                 704
```

Fig. 8.5. Alterations in subroutine GETCOR

The subroutines ROTAT, TRANSL, CARTES and POLAR also listed in Chap. 7 are independent of the coordinate system used and remain the same. Of the subroutines for input/output, which can be found on the diskette but have not been discussed, one only needs to change ZEICH, which outputs the positions of the two masses. In the main program KAP8 and in GETCOR one also needs to modify some text, in which the earth and the moon are explicitly mentioned.

The complete program is to be found under the title KAP-08.FOR on the diskette.

8.4 Exercises

8.4.1 Study the behaviour of the small mass μ in the vicinity of the Lagrange points. Verify that configurations in which the masses m_1, m_2 and μ lie on a line are unstable, and that the triangular configurations are stable. Investigate the size and shape of the zone of stability around L_4. *Hint:* Choose coordinate system 4 or 5 for the exercise.

8.4.2 Try to find a horseshoe orbit. *Hint:* Choose m_2/m_1 very small, e.g. $m_2/m_1 = 0.001$. Use coordinate system 4 for the exercise and place the mass μ in the vicinity of the Lagrange point L_3, see Fig. 8.1.

8.4.3 Investigate how the orbit of the small mass μ is perturbed by m_2, when they have orbital periods in the ratio 1:2 (or 1:3). Again use a very small ratio m_2/m_1.

8.5 Solutions to the Exercises

8.5.1 The stable zone about L_4 is greater, the smaller m_2/m_1 is. The extent of the zone in the φ direction is somewhat greater than in the r direction. This is not surprising, since according to Kepler's third law a change in the radius leads, after all, to a change in the orbital velocity.

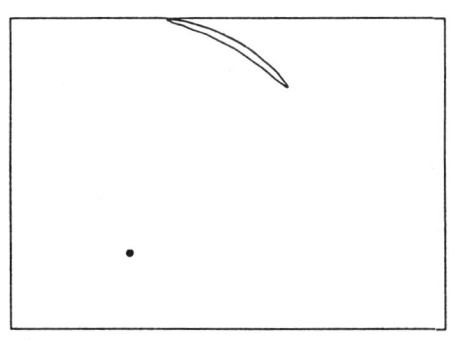

Fig. 8.6. Libration oscillations about L_4

Fig. 8.7. Horseshoe orbit

8.5.2 The horseshoe orbit shown in Fig. 8.7 is obtained with

$m_2/m_1 = 0.001,$
$r_d = 384\,000 \text{ km},$
$\varphi_d = 180°,$
$v_d = 0,$
$\psi_d = 0°$

in coordinate system 4.

8.5.3 Using the input parameters

$m_2/m_1 = 0.001,$
$r_d = 241\,900 \text{ km},$
$\varphi_d = 90°,$
$v_d = 4680 \text{ km/h},$
$\psi_d = 180°$

in coordinate system 1, one first obtains a nearly circular orbit, whose period is very near 324 h, and so half the period of m_2. After about 11 000 h the orbit has become very much more eccentric; r_d now oscillates between 209 000 and 268 000 km. After 23 000 h the orbit is again nearly circular with an average orbital radius of 242 000 km.

9. Computation of Electric Fields by the Method of Successive Over-relaxation

9.1 Formulation of the Problem

With the methods described in Chaps. 7 and 8 we can also solve the equations of motion for charged particles in electromagnetic fields. In this way one can, for example, study the focussing properties of guide fields in particle accelerators or the imaging properties of lenses in electron optics. The assumption here is that the electromagnetic fields are known, and this brings us to the next problem, namely, the boundary problem in partial differential equations (in contrast to the initial value problems considered up to now).

Electrostatics and magnetostatics are usually presented in classes so that first Maxwell's equations are derived and then, by means of simple examples, one is introduced to the form of their solutions. The examples include the plate condenser, the spherical condenser, the point charge in front of the conducting plane, the infinite coil, the field of a straight line conductor bearing a current, etc... A feeling for inhomogeneous fields is also obtained from the well known sheet of paper covered with iron filings, and with a magnet underneath. One's imagination, however, is strained in passing from that to understanding the penetration factor of the control grid of an electronic valve, or picturing the field configuration in the structural elements of electron optics. In what follows it will now be shown how one can, by means of the computer, enlarge the repertoire of examples with well known field patterns.

We restrict ourselves to electric fields in the vacuum, i.e. we study solutions of Laplace's equations,

$$\Delta \Phi = 0 \ . \tag{9.1}$$

As boundary condition we specify the potential on the boundary of the region in which the equation is to be solved.

We choose as an example a simple electrostatic lens which, with slight modifications, is still used in electron optics. A cylindrical metal tube with radius r_0 is interrupted by flanges F_1, F_2, as shown in Fig. 9.1.

The flanges form a plate condenser with the two annular plates F_1, F_2 at a separation $2d$, to which the two pieces of tube R_1, R_2 are attached. Let the right-hand flange be charged to a potential of 1000 V, and the left to a potential of -1000 V.[1] One seeks the form of the potential $\Phi(x, y, z)$ in the

---
[1] In electron optics higher voltages are usually employed. To avoid field emission the corners of the tube/flange transitions have to be rounded off; this is true even at 1000 V for very small values of d. In our calculation we shall not take account of this rounding off of the corners.

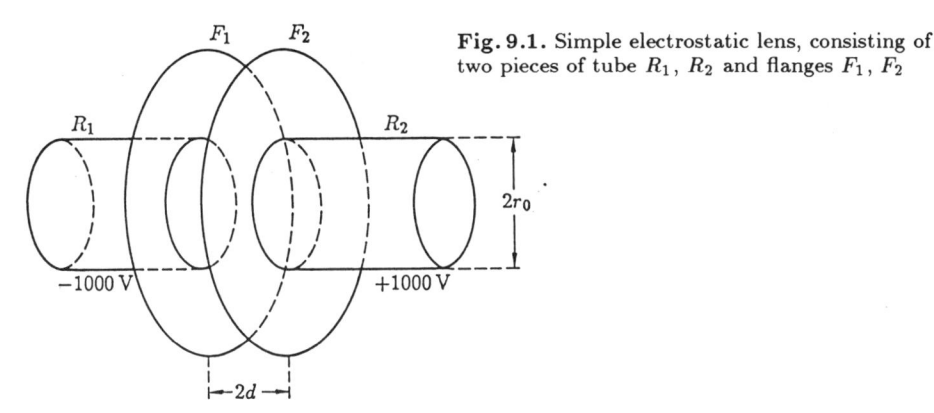

Fig. 9.1. Simple electrostatic lens, consisting of two pieces of tube R_1, R_2 and flanges F_1, F_2

region between the plates and in the tubes. If the potential is known as a function of the position coordinates, then one obtains the electric field from

$$\boldsymbol{E} = -\text{grad}\,\Phi \quad . \tag{9.2}$$

The solution of Lapace's equation (9.1) is uniquely determined in a closed volume, when the potential is specified over the surface of the volume. Up to now the volume in which we wish to compute the potential function is still open. Using a little physical intuition, however, we can close it. If we go far enough along the tube, then the potential in each tube is scarcely going to change any more. This means that, to the far right and the far left of the flanges we shall find $\Phi_1 = +1000\,\text{V}$ and $\Phi_2 = -\Phi_1 = -1000\,\text{V}$, respectively, in the tubes. In other words, the potential field will not be changed if we close the tubes far to the right and far to the left of the flanges with metal lids. Between the plates the potential field will be more and more similar to that of an ordinary plate condenser, the further we are from the axis of the tubes. This means that here also we can close the volume, only this time not with a metal plate, buth with an insulator or simply with an imaginary surface. On this surface the potential varies linearly with distance from the plates, from $+1000\,\text{V}$ on the right-hand plate to $-1000\,\text{V}$ on the left-hand plate. The potential is thereby specified on the surface of a closed volume, and we can now consider how to solve Laplace's equation.

First we make use of the axial symmetry of the configuration in order to reduce the number of coordinates from 3 to 2. We introduce cylindrical coordinates. The axis of the tubes is the z-axis. The coordinates x, y are replaced by plane polar coordinates r, φ, where r denotes the distance from the z-axis and φ the azimuthal angle around the z-axis. In cylindrical coordinates Laplace's equation becomes:

$$\Delta\Phi = \frac{\partial^2\Phi}{\partial z^2} + \frac{1}{r}\frac{\partial\Phi}{\partial r} + \frac{\partial^2\Phi}{\partial r^2} + \frac{1}{r^2}\frac{\partial^2\Phi}{\partial \varphi^2} = 0 \quad , \quad (r > 0) \quad . \tag{9.3}$$

Because of the axial symmetry the potential Φ cannot depend on the azimuthal angle φ, and the equation simplifies to

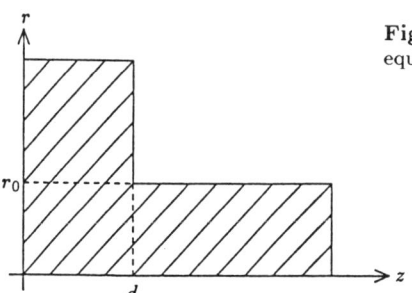

Fig. 9.2. The region in which Laplace's equation is to be solved

$$\Delta \Phi = \left(\frac{\partial^2}{\partial z^2} + \frac{1}{r}\frac{\partial}{\partial r} + \frac{\partial^2}{\partial r^2} \right) \Phi(z,r) = 0 \quad , \quad (r > 0) \quad . \tag{9.4}$$

Moreover, we see that the configuration is symmetric about the plane $z = 0$. It is therefore sufficient to solve (9.4) in the shaded region of Fig. 9.2. This region is still open below. However, since the lower edge is identical with the axis of symmetry, no boundary condition is needed here.

For the numerical solution of (9.4) with the boundary values specified, we use a finite element method with successive over-relaxation.

9.2 Numerical Method

9.2.1 Discretisation of Laplace's Equation

For the numerical solution of Laplace's equation we shall approximate $\Phi(z,r)$ on a two-dimensional grid of points with constant mesh width h. Instead of the function $\Phi(z,r)$ we shall therefore consider only a matrix $(\Phi_{i,k})$ of function values. Let

$$z_i = ih \quad , \quad r_k = kh \quad , \quad \Phi_{i,k} = \Phi(z_i, r_k) \quad . \tag{9.5}$$

Since we have chosen a very simple shape for the electrode boundaries, it makes it easy to fit these into the grid. We only have to choose h so that the tube radius r_0 and the semi-distance separating the flanges d are integral mulitples of h. More complicated shapes of electrode have to be represented by a polygonal approximation.

In order to solve Laplace's equation (9.4) on this grid, we have first to discretise the differential operators $\partial/\partial r$, $\partial^2/\partial r^2$ and $\partial^2/\partial z^2$. We use the formulae (2.6) and (2.8) and obtain for $r > 0$:

$$\begin{aligned}
\Delta \Phi &= \left(\frac{\partial^2}{\partial z^2} + \frac{\partial^2}{\partial r^2} + \frac{1}{r}\frac{\partial}{\partial r} \right) \Phi(z,r) \\
&= \frac{1}{h^2}(\Phi(z,r+h) + \Phi(z,r-h) + \Phi(z+h,r) + \Phi(z-h,r) - 4\Phi(z,r)) \\
&\quad + \frac{1}{r}\frac{1}{2h}(\Phi(z,r+h) - \Phi(z,r-h)) + O(h^2) = 0 \quad .
\end{aligned} \tag{9.6}$$

We now apply the relations (9.5), neglect the error term $O(h^2)$, multiply the equation by h^2 and so obtain an approximation of Laplace's equation on the grid, for $k > 0$:

$$(\Phi_{i,k+1} + \Phi_{i,k-1} + \Phi_{i+1,k} + \Phi_{i-1,k} - 4\Phi_{i,k}) + \frac{1}{2k}(\Phi_{i,k+1} - \Phi_{i,k-1}) = 0 \ . \tag{9.7}$$

For $k = 0$, i.e. for the grid points on the axis of symmetry, (9.7) is not valid. For these points we have to formulate a special equation. The equation is most simply obtained by first converting back into Cartesian coordinates x, y, z and using (2.8):

$$\Delta \Phi(x,y,z) = \frac{1}{h^2}(\Phi(x+h,y,z) + \Phi(x-h,y,z) + \Phi(x,y+h,z)$$
$$+ \Phi(x,y-h,z) + \Phi(x,y,z+h) + \Phi(x,y,z-h)$$
$$- 6\Phi(x,y,z)) + O(h^2) = 0 \ . \tag{9.8}$$

We wish to apply this equation at the axis of symmetry and accordingly set $x = y = 0$. Now we return from the potential function $\Phi(x,y,z)$ in Cartesian coordinates to the potential function $\Phi(z,r)$ in cylindrical coordinates. The first four function values on the right-hand side are all equal, namely equal to $\Phi(z,h)$. If we again neglect $O(h^2)$, multiply throughout by h^2 and use the relations (9.5), we obtain the so-called axis formula:

$$4\Phi_{i,1} + \Phi_{i+1,0} + \Phi_{i-1,0} - 6\Phi_{i,0} = 0 \ . \tag{9.9}$$

9.2.2 The Method of Successive Over-relaxation

If one writes down (9.7) for all interior points of the grid and (9.9) for all points on the axis of symmetry (with the exception of the upper and lower boundary points), then one obtains a system of linear equations, from which in principle the matrix $(\Phi_{i,k})$ can be calculated. Since we are to calculate many function values $\Phi_{i,k}$, however, this system of equations has a high dimension. The usual methods for solving systems of linear equations, such as the Gaussian elimination method, are then no longer suitable for the solution of the problem.

The system of equations (9.7), (9.9), however, has one peculiarity: each point is linked only with the directly neighbouring points. If we represent the Laplace operator as an enormous matrix, this matrix is almost empty: most of the elements are zero. For such problems the method of successive over-relaxation [9.1] has been developed.

The system of linear equations is solved by iteration:

1. One postulates an approximate solution.
2. From this one calculates by means of a formula (characterising the method) a better approximate solution.
3. One tests whether the improved approximate solution fulfils a quality criterion. If this is not so, one replaces the original approximate solution by the improved approximate solution and returns to stage 2.

Now how do the individual steps of the method look?

For 1: For the unknown potential values $\Phi_{i,k}$ in the interior region of the grid and on the axis of symmetry, one postulates values which one feels to be reasonable. The method of successive over-relaxation converges even if one postulates a constant value for Φ; by judicious choice one merely economises in iteration steps.

For 2: For each point (i,k), one after the other, in the interior and on the axis of symmetry, one applies (9.7) and (9.9), to calculate the value of a potential U at the point (i,k) from the potential values at the neighbouring points,

$$U = \begin{cases} \frac{1}{4}(\Phi_{i,k+1} + \Phi_{i,k-1} + \Phi_{i+1,k} + \Phi_{i-1,k}) \\ \qquad + \frac{1}{8k}(\Phi_{i,k+1} - \Phi_{i,k-1}) \;,\; & k>0 \\ \frac{1}{6}(\Phi_{i+1,k} + \Phi_{i-1,k} + 4\Phi_{i,k+1}) \;,\; & k=0 \end{cases} \quad (9.10)$$

The improved approximate solution at the point (i,k) is calculated by means of the formula

$$\Phi_{i,k}^{\text{new}} = \Phi_{i,k}^{\text{old}} + w(U - \Phi_{i,k}^{\text{old}}) \;,\; 1 \leq w < 2 \;. \quad (9.11)$$

The constant w is chosen so that one obtains the fastest possible convergence. We shall consider this in more detail. The $\Phi_{i,k}$ are represented by merely one array of numbers in the memory. Whenever a new value $\Phi_{i,k}$ is calculated, the old value is overwritten by the new. The order in which (9.11) is applied to the points of the grid is arbitrary.

For 3: When the elements of the matrix $(\Phi_{i,k})$ are no longer changing, i.e. when

$$\Phi_{i,k}^{\text{new}} = \Phi_{i,k}^{\text{old}} \;, \quad (9.12)$$

then the matrix of the potential values $\Phi_{i,k}$ satisfies the discretised Laplace's equation with the specified boundary condition. The quality or convergence criterion is that at no point (i,k) does $\Phi_{i,k}^{\text{new}}$ differ from $\Phi_{i,k}^{\text{old}}$ by more than a specified small value ε. In our example we shall specify for ε the value $\varepsilon = 10^{-7}\Phi_1 = 10^{-4}$ V.

A peculiarity of the method is the parameter w in (9.11). If one sets $w = 1$, then one simply replaces $\Phi_{i,k}^{\text{old}}$ by U. From (9.10) one can see that U is calculated only from the potential values at the neighbouring grid points. At each application of (9.11), therefore, the information originating from the boundary values can only propagate by one mesh width. The propagation of the information has a certain similarity to a diffusion process. With $w = 1$ the approximate values $\Phi_{i,k}$ move slowly and monotonically in the course of the iteration towards the respective end values. This process can be accelerated by setting $w > 1$. This is known as over-relaxation, in contrast to a relaxation with $w \leq 1$. If too large a value is chosen for w, then oscillation and instability set in. It is known that $w \geq 2$ always leads to instability in the method shown here. A suitable value for w is obtained by trial. It is difficult to predict theoretically the most suitable value. However, it is known that for too large values of w the convergence rapidly becomes worse, so it is better to begin with smaller values of w.

9.3 Programming

First of all we replace the semi-separation distance of the flanges d and the tube radius r_0 by the dimensionless integers m and n,

$$m = \frac{d}{h} \quad , \quad n = \frac{r_0}{h} \quad . \tag{9.13}$$

The assumed wall which closes the end of the tube is at a distance $m_{\max}$ from the plane $z = 0$, and the wall which encloses the flanges is at a distance $n_{\max}$ from the axis of symmetry. We have thus chosen h as the unit of length and eliminated it from the calculation. We specify $m_{\max}$ and $n_{\max}$ as fixed and leave m and n variable. In practical examples with fine grids, in order to save computation time, one starts with rather small values for $m_{\max}$ and $n_{\max}$. The solution will then show whether these values are too small and must be increased. In our program we shall specify $m_{\max}$ and $n_{\max}$ sufficiently large from the start. In Fig. 9.3 we see a sketch of the grid of points.

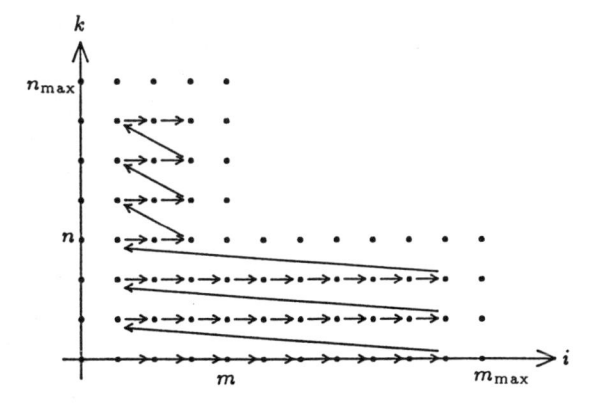

Fig. 9.3. Grid of points for the region shown in Fig. 9.2; the arrows show the order in which the points of the grid are processed in each step of the iteration

The method of successive over-relaxation can be programmed with little effort. We carry out the whole computation in the main program KAP9 (see Fig. 9.5; the notation is shown in Fig. 9.4). The values for $m_{\max}$ and $n_{\max}$ as well as for Φ_1 and ε are specified in the PARAMETER instruction in line 102; we input the values for m, n and w, as well as a potential value Φ_0, which we need for setting up a zero order approximate solution. At the start of the iteration we specify the following approximate solution (see Fig. 9.3): for the interior of the tube (region from $i = m$ to $i = m_{\max}$) we choose a constant potential Φ_0

Physical title	FORTRAN title	Physical title	FORTRAN title
m	M	Φ_0	PHI0
n	N	Φ_1	PHI1
m_{max}	MMAX	w	W
n_{max}	NMAX	ε	EPS
i	I	U	U
k	K	$\Phi_{i,k}^{old}$	PHIOLD
		$\Phi_{i,k}, \Phi_{i,k}^{new}$	PHI(I,K)

Fig. 9.4. Notation used in the main program KAP9

```
          PROGRAM KAP9                                                  100

          IMPLICIT DOUBLE PRECISION (A-H,O-Z)                           101
          PARAMETER (NMAX=50, MMAX=2*NMAX, PHI1=1000.D0, EPS=1.D-7*PHI1) 102
          DIMENSION PHI(0:MMAX,0:NMAX)                                  103
*         (Input: M,N,PHI0,W)

          ITERA=1                                                       104
          DO 20 K=0,NMAX-1                                              105
            DO 10 I=1,M-1                                               106
              PHI(I,K)=(I*PHI0)/M                                       107
10        CONTINUE                                                      108
20        CONTINUE                                                      109
          DO 40 K=0,N-1                                                 110
            DO 30 I=M,MMAX-1                                            111
              PHI(I,K)=PHI0                                             112
30        CONTINUE                                                      113
40        CONTINUE                                                      114
          DO 50 K=0,NMAX                                                115
            PHI(0,K)=0.D0                                               116
50        CONTINUE                                                      117
          DO 60 I=1,M-1                                                 118
            PHI(I,NMAX)=(I*PHI1)/M                                      119
60        CONTINUE                                                      120
          DO 70 K=N,NMAX                                                121
            PHI(M,K)=PHI1                                               122
70        CONTINUE                                                      123
          DO 80 I=M+1,MMAX                                              124
            PHI(I,N)=PHI1                                               125
80        CONTINUE                                                      126
          DO 90 K=0,N-1                                                 127
            PHI(MMAX,K)=PHI1                                            128
90        CONTINUE                                                      129
100       CONTINUE                                                      130
          EPSMAX=0.D0                                                   131
          DO 120 K=0,NMAX-1                                             132
            IF(K.LT.N) THEN                                             133
              MLIM=MMAX-1                                               134
            ELSE                                                        135
              MLIM=M-1                                                  136
            ENDIF                                                       137
            DO 110 I=1,MLIM                                             138
              IF(K.EQ.0) THEN                                           139
```

```
              U=(PHI(I+1,K)+PHI(I-1,K)+4.D0*PHI(I,K+1))/6.D0              140
              ELSE                                                         141
              U= (PHI(I,K+1)+PHI(I,K-1)+PHI(I+1,K)+PHI(I-1,K))/4.D0        142
        &       +(PHI(I,K+1)-PHI(I,K-1))/(8.D0*K)                          143
              ENDIF                                                        144
              PHIOLD=PHI(I,K)                                              145
              PHI(I,K)=PHIOLD+W*(U-PHIOLD)                                 146
              IF(ABS(PHI(I,K)-PHIOLD).GT.ABS(EPSMAX)) EPSMAX=PHI(I,K)-PHIOLD  147
  110       CONTINUE                                                       148
  120     CONTINUE                                                         149

  *       (Output: ITERA,EPSMAX)

          ITERA=ITERA+1                                                    150
          IF(ABS(EPSMAX).GT.EPS) GOTO 100                                  151

  *       (Output: PHI)

          END                                                              152
```

Fig. 9.5. Numerical portion of the main program KAP9

and in the space between the plates (region from $i = 0$ to $i = m$) we let the potential increase linearly with i from 0 to Φ_0. In the two DO-loops from lines 105 to 114 this preliminary definition is carried out. The following DO-loops from lines 115 to 129 then serve to assign the potential distribution of the boundary values.

In the IF-block of lines 139 to 144 U is computed from (9.10). Subsequently, in lines 145 and 146 the new potential values are determined from (9.11). The whole is then enclosed in two nested DO-loops, of which the inner loop runs through the index i (beginning with line 138), and the outer loop the index k (beginning with line 132). In line 147 the maximal potential change in each iteration is obtained and assigned to the variable EPSMAX. After each iteration a test is made to see whether the maximal potential change is still greater than ε (line 151). If this is the case, we carry out a further iteration, otherwise $\Phi_{i,k}$ is output.

We provide in the program for a graphical and a numerical output for Φ. In the numerical output, for reasons of space, only the quadrant shown in Fig. 9.3 is printed. The values for $\Phi_{i,k}$ are arranged in chess-board fashion, with the index k plotted in the vertical direction, and the index i in the horizontal direction. We recall once again that i runs along the z-axis and k the r-axis. In Figs. 9.6 to 9.8 we see the output for our above example.

For the graphical output we give not only one quadrant of the lens, but the whole lens in the r, z plane. Equipotential lines are shown at intervals of 100 V. In Fig. 9.9 we see the result for $m = 4$, $n = 8$, $\Phi_0 = 1000$ V and $w = 1.5$; this needed 72 iterations

To draw the equipotential lines we use the subroutines GMOVE and GDRAW. First we calculate individual points on the equipotential lines, and then join them with straight lines by means of GDRAW. The location of points of equal potential is explained with the help of Fig. 9.10. Here we see a detail from

48	0	750												
45	0	750												
42	0	750												
39	0	750												
36	0	750												
33	0	750												
30	0	750												
27	0	750												
24	0	750												
21	0	750												
18	0	750												
15	0	750												
12	0	749												
9	0	733												
6	0	557	865	951	981	992	997	999	999	1000	1000	1000	1000	1000
3	0	440	743	891	955	982	992	997	999	999	1000	1000	1000	1000
0	0	409	707	870	945	977	991	996	998	999	1000	1000	1000	1000
K/I	0	3	6	9	12	15	18	21	24	27	30	33	36	39

Fig. 9.6. Numerical output for $m = 4$, $n = 8$ (in steps of $3h$)

40	0	500	1000											
38	0	500	1000											
36	0	500	1000											
34	0	500	1000											
32	0	500	1000											
30	0	500	1000											
28	0	500	1000											
26	0	500	1000											
24	0	500	1000											
22	0	500	1000											
20	0	500	1000											
18	0	500	1000											
16	0	500	1000											
14	0	500	1000											
12	0	498	1000											
10	0	491	1000											
8	0	460	1000	1000	1000	1000	1000	1000	1000	1000	1000	1000	1000	1000
6	0	381	706	865	932	964	981	990	994	997	998	999	999	1000
4	0	322	592	772	875	932	962	979	989	994	997	998	999	999
2	0	290	540	722	840	910	950	972	985	992	995	997	999	999
0	0	281	524	707	828	902	945	970	983	991	995	997	998	999
K/I	0	2	4	6	8	10	12	14	16	18	20	22	24	26

Fig. 9.7. Numerical output for $m = 4$, $n = 8$ (in steps of $2h$)

Fig. 9.8, with grid points marked by a solid circle and points on the 400 V equipotential line by squares.

The point grid is covered from right to left and from bottom to top, in contrast to Fig. 9.3. If the potential value 400 V lies between the potential values of two neighbouring points, linear interpolation is used to determine the point at which the potential amounts to 400 V. In Fig. 9.10 the first 400 V point is found between the grid points with 505 and 348 V, then the point between the two grid points at 557 and 381 V. The next point found is that between 381

K\I	0	1	2	3	4	5	6	7	8	9	10	11	12	13
20	0	250	500	750	1000									
19	0	250	500	750	1000									
18	0	250	500	750	1000									
17	0	250	500	750	1000									
16	0	250	500	750	1000									
15	0	250	500	750	1000									
14	0	250	500	750	1000									
13	0	249	499	749	1000									
12	0	249	498	749	1000									
11	0	247	496	747	1000									
10	0	244	491	743	1000									
9	0	239	481	733	1000									
8	0	228	460	707	1000	1000	1000	1000	1000	1000	1000	1000	1000	1000
7	0	211	421	627	811	889	929	952	966	976	982	987	991	993
6	0	193	381	557	706	803	865	905	932	951	964	974	981	986
5	0	177	348	505	639	740	813	864	901	928	947	961	971	979
4	0	164	322	467	592	694	772	831	875	908	932	949	962	972
3	0	154	303	440	561	661	743	806	855	891	919	940	955	967
2	0	148	290	422	540	640	722	788	840	879	910	933	950	963
1	0	144	283	412	528	627	710	778	831	872	904	928	946	960
0	0	143	281	409	524	623	707	774	828	870	902	927	945	959

Fig. 9.8. Numerical output for $m = 4$, $n = 8$ (in steps of h)

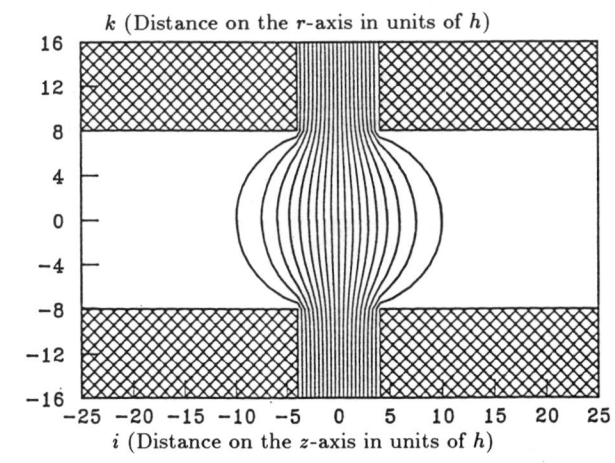

Fig. 9.9. Graphical output for $m = 4$, $n = 8$

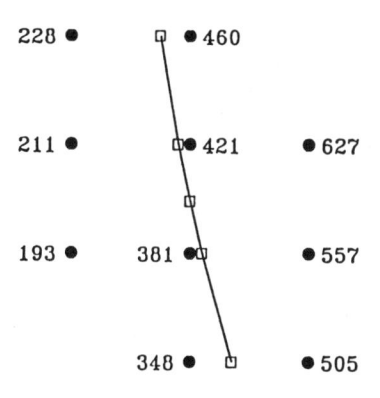

Fig. 9.10. A few grid points with their potential values and the 400 V equipotential line for $m = 4$, $n = 8$

99

and 421 V, and so on. Each newly found point on the 400 V equipotential line is joined by a straight line to the previous point found. By reflection in the axes we obtain the equipotential lines for the whole lens.

In order to make the interior of the lens stand out, we have hatched the outer region in Fig. 9.9 by means of the subroutine GBOX. The call

CALL GBOX(X1, Y1, X2, Y2, ICOLOR, ISTYLE)

has the effect that a rectangle is drawn, with the bottom left-hand corner defined by the coordinates (X1,Y1), and the top right-hand corner by the coordinates (X2,Y2). The variable ICOLOR indicates the colour (see discussion of GLINE, GMOVE and GCOLOR in Chap. 4), the variable ISTYLE the density of infilling of the rectangle. The following values can be used for ISTYLE: 1 = empty, 2 = solid, 3 = densely shaded, 4 = moderately shaded, 5 = lightly shaded, 6 = densely cross-hatched, 7 = moderately cross-hatched, 8 = lightly cross-hatched.

The complete program presented here in summary is to be found on the diskette in the file KAP-09.FOR.

9.4 Exercises

9.4.1 Find by experiment the most favourable over-relaxation factor w. What happens when $w = 1$, and when $w \to 2$?

9.4.2 The smallness of the convergence limit ε does not necessarily guarantee a high accuracy of the solution obtained. Can one, using the above program, investigate how good the solution actually is, i.e. how well it approximates to the true solution of Laplace's equation?

9.4.3 Calculate the potential field for different lens parameters, and consider qualitatively how to trace the paths of electrons. How does a Wehnelt cylinder work? (*Comment*: A Wehnelt cylinder encloses a thermionic cathode and converges the paths of the electrons so that practically all the electrons pass through a point).

Solutions to the Exercises

9.5.1 In order to achieve a rather short computation time we set $m = n = 5$, $\Phi_0 = 1000$ V and try $w = 1.0, 1.2, 1.4, 1.6, 1.8, 1.9$ and 1.99.

With $w = 1.0$ we have slow convergence. The largest change of a potential value on the grid decreases monotonically from iteration to iteration. The series of largest changes is similar to a geometrical series, i.e. with each iteration the largest change decreases by several per cent. Qualitatively the same behaviour is found with $w = 1.2$ and $w = 1.4$; of course, the convergence is now faster With $w = 1.0$ one needs 99 iterations to achieve a convergence limit of $\varepsilon = 0.0001$ V. With $w = 1.2$ it is still 67 iterations, and with 1.4 only 40.

With $w = 1.6$ the iteration converges still faster; now only 18 steps are needed. The series of largest changes, however, no longer resembles a geometric series. The changes start to jump about. With still larger values of w the jumps become stronger, and the convergence is poorer. With $w = 1.8$ one needs 58 iterations, and with $w = 1.9$ one needs 128. With $w = 1.99$ the jumps dominate. After 500 iterations the changes are smaller than 1 V. The convergence limit is reached after 1339 iterations.

For other values of m and n one finds a similar behaviour. If one wishes to specify a standard value for w, this should lie at about $w = 1.5$ or 1.6.

The standard value for Φ_0 is 1000 V. The method also converges with other values such as, for example, with $\Phi_0 = 0$ V. It then costs a few preliminary iterations, before the approximation achieves the quality which it had from the start with $\Phi_0 = 1000$ V. From then on the behaviour of the iterations is little different.

9.5.2 The form of the solution of Laplace's equation, i.e. the shape of the equipotential surfaces, cannot depend on the choice of the units for length and potential. In our program we have used the quantity h, not only as the mesh width, but also as the unit of length. Accordingly we obtain the same solution with specified parameters m and n for all values of h. If we wish to calculate the equipotential surfaces for a physically specified electron lens with grids of two different degrees of fineness, in order thereby to draw a conclusion on the accuracy of the solution, we must change m and n, without changing the ratio m/n.

To test the accuracy of the method we specify the lens parameters by the choice of $m = 5$ and $n = 3$, and then carry out the computation with $m = 5$, $n = 3$, with $m = 10$, $n = 6$ and with $m = 20$, $n = 12$. We compare the potential values $\Phi_{i,k}$ of the first computation with the potential values $\Phi_{2i,2k}$ of the second computation and the potential values $\Phi_{4i,4k}$ of the third computation.

We find that with the first doubling of the number of mesh points (per dimension) the potential values change by at most about 1%; with the second doubling only by at most 0.2%. One may conclude from this that $m = 5$, $n = 3$ gives a satisfactorily fine grid. In the graphical output there is an additional inaccuracy from the linear interpolation between the grid points. With grids finer than $m = 5$, $n = 3$ the total error will nevertheless be smaller than the thickness of the lines drawn on the screen.

9.5.3 The equipotential lines of an electron lens are to be seen in Fig. 9.9. Qualitative consideration shows that one half of the lens (relative to $z = 0$) always acts as a dispersing lens and the other half as a converging lens. However, since the electrons do not pass through the halves of the lens with the same velocity, the characteristics of the two do not cancel out. The characteristics of the converging lens dominate.

Another interesting case is that of a "thick" lens. Figure 9.11 shows the equipotential lines for $m = 20$, $n = 8$. We ignore the right-hand half of the lens and then have, where the left-hand tube penetrates the flange, a similar field pattern to that in the neighbourhood of the hole in the Wehnelt cylinder. Slow electrons, coming from various directions on the left, are attracted by the anode potential and accelerated towards the right. At the same time they are concentrated towards the z-axis. All the electrons pass through a region in the neighbourhood of the axis which is smaller than the cross-section of the opening from which the electrons come. In this way one obtains an almost point-like electron source, which can be imaged by additional electron-optical elements (e.g. on to the screen of a personal computer!).

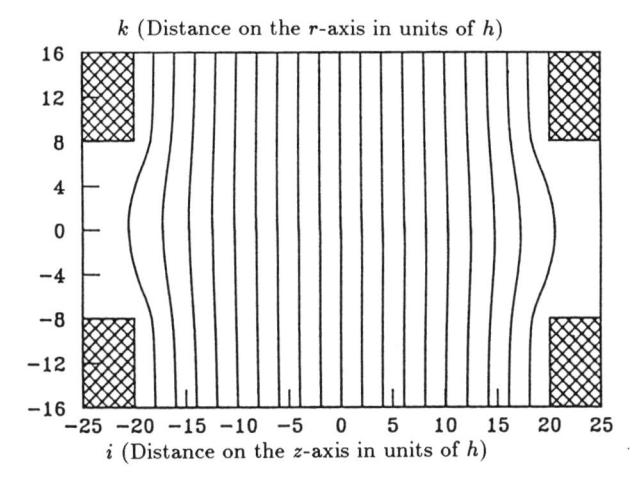

Fig. 9.11. Behaviour of the equipotential lines for $m = 20$, $n = 8$

10. The Van der Waals Equation

10.1 Formulation of the Problem

In this chapter we shall show how a computer can help us to gain a deeper understanding of a simple equation of physics and to test the agreement between theory and experiment. We consider the van der Waals gas equation. It is obtained from the ideal gas equation by including corrections for real gases.

It is well known that the equation of state for ideal gases describes a gas of point-like particles in the limit of vanishing mutual interaction. The gas particles interact with the container walls only by the transfer of momentum during reflection. The ideal gas equation for one mole of gas is:

$$PV = RT \quad , \tag{10.1}$$

where P is the pressure exerted by the gas on the container walls, V the volume of one mole, R the gas constant and T the absolute temperature. The ideal gas equation is a good approximation for inert gases at room temperature and atmospheric pressure, and under these conditions it also applies quite well to air. On the other hand it is no longer valid for butane or carbon dioxide compressed in steel flasks.

Van der Waals adapted the ideal gas equation to real gases by the following changes [10.1]:

1. Real gas particles have a finite size. When packed as densely as possible they occupy the so-called eigenvolume. The steep rise in pressure shown by the ideal gas equation as $V \to 0$ must occur in reality when the gas is compressed to its eigenvolume. This is achieved by replacing the volume V in the ideal gas equation by $(V - b)$, where b is the total volume of the particles in one mole.
2. Beyond the short-range region of repulsion there is an attractive force between the gas particles, which is today known as the van der Waals force. It can be calculated in quantum mechanical models as an interaction of induced electric dipoles which decreases with a high power of the distance between the particles. In the gas equation this attractive force has the effect that the pressure exerted by the container walls on the gas is augmented by an internal pressure which helps the external pressure to compress the gas. Van der Waals takes account of this internal pressure by the term a/V^2.

If one inserts the two corrections in the ideal gas equation one obtains the van der Waals equation:

$$\left(P + \frac{a}{V^2}\right)(V - b) = RT \quad . \tag{10.2}$$

The isotherms of this equation, i.e. the curves of equal temperature in the P-V diagram, have the appearance shown in Fig. 10.1. For temperatures below a critical temperature T_c one has three values of V for each value of the pressure. This at first appears physically unrealistic. Real gases behave differently. If one decreases the volume at a constant temperature the pressure increases until the vapour pressure of the liquid phase is reached. Then condensation follows, and the pressure no longer increases, until all the gas is liquefied. For further decrease in the volume, the pressure then climbs very steeply.

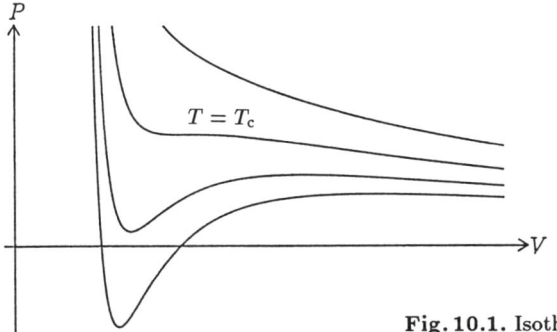

Fig. 10.1. Isotherms of the van der Waals equation

The van der Waals equation does not contain the phenomenon of condensation. However it is not physically unrealistic. At least in the immediate neighbourhood of the values of P, V and T where condensation would occur, it still describes rather well the formation of the super-saturated vapour. The incipient condensation in the presence of condensation nuclei must be incorporated into the van der Waals equation by means of an additional prescription. Maxwell has done this by introducing the lines named after him. For each isotherm below the critical temperature Maxwell defines a horizontal line in the P-V diagram, such that the two shaded areas in Fig. 10.2 are of equal size. This can also be formulated as follows: The area of the rectangle below the

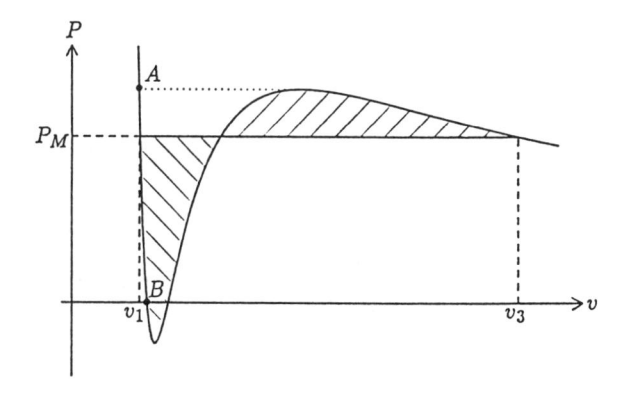

Fig. 10.2. Isotherm of the van der Waals equation and the Maxwell line

Maxwell line must be as large as the integral under the isotherm in the same interval. Then, whenever the gas phase and the liquid phase co-exist, the state of the real gas lies on the Maxwell line and not on the van der Waals isotherm.

As already mentioned, van der Waals adjusted the ideal gas equation to real gases using only two parameters a and b. It was not to be expected that this simple correction would lead to complete agreement between theory and experiment for all real gases. Especially in the overlapping region between repulsion and attraction of the gas particles, the two parameters cannot describe the details of the interaction. With the computer and the repertoire of programs already constructed for numerical and graphical output we can display the isotherms of the van der Waals equation without much effort. The inclusion of the Maxwell lines requires only little computation time. Once one has determined the Maxwell lines for a number of temperatures one obtains a theoretical vapour pressure curve and can compare it with a vapour pressure curve obtained experimentally.

10.2 Numerical Method

The van der Waals equation determines a point in the P-V diagram where the two zeros of the derivative $\partial P/\partial V$ on an isotherm coincide, and as a result the second derivative $\partial^2 P/\partial V^2$ is also zero; see Fig. 10.1. This point is called the critical point. The associated physical quantities are the critical pressure P_c, the critical temperature T_c and the critical volume V_c. We obtain the critical point from the conditions:

$$\frac{\partial P}{\partial V} = \frac{-RT_k}{(V_k - b)^2} + \frac{2a}{V_k^3} = 0 \quad,\quad \text{whence} \quad \frac{RT_k}{(V_k - b)^2} = \frac{2a}{V_k^3} \qquad (10.3a)$$

and

$$\frac{\partial^2 P}{\partial V^2} = \frac{2RT_k}{(V_k - b)^3} - \frac{6a}{V_k^4} = 0 \quad,\quad \text{whence} \quad \frac{2RT_k}{(V_k - b)^3} = \frac{6a}{V_k^4} \quad. \qquad (10.3b)$$

On dividing (10.3a) by (10.3b) we obtain

$$\frac{V_k - b}{2} = \frac{V_k}{3} \quad,\quad V_k = 3b \quad, \qquad (10.4a)$$

and after a little algebra:

$$T_k = \frac{8a}{27Rb} \quad, \qquad (10.4b)$$

$$P_k = \frac{a}{27b^2} \qquad (10.4c)$$

$$P_k V_k = \tfrac{3}{8} RT_k \quad, \qquad (10.4d)$$

from which we see that the gas pressure at the critical point is $\tfrac{3}{8}$ times as great as that of the ideal gas at the same density and temperature.

An abbreviated way of writing the van der Waals equation is obtained by introducing the reduced (dimensionless) quantities $p = P/P_c$, $v = V/V_c$ and $t = T/T_c$:

$$\left(\frac{pa}{27b^2} + \frac{a}{9b^2v^2}\right)(3bv - b) = \frac{8at}{27b} \quad,$$

$$\left(p + \frac{3}{v^2}\right)(3v - 1) = 8t \quad. \tag{10.5}$$

This makes it clear that the P-V diagrams for all van der Waals gases are similar and differ from each other only by scale factors. It follows from (10.5) that the reduced eigenvolume (volume of the particles) is just $\frac{1}{3}$.

The condition for the Maxwell lines is the same in the p-v diagram as in the P-V diagram: the integral under the isotherm must be equal to the area under the Maxwell line. Denoting the end points of the Maxwell line by v_1 and v_3, and the vapour pressure by p_M, the condition is

$$p_M(v_3 - v_1) = \int_{v_1}^{v_3} p(v)dv \quad. \tag{10.6}$$

In order to evaluate this relation, we first eliminate p_M using (10.5). For the left-hand side of (10.6) we obtain:

$$p_M(v_3 - v_1) = \left(\frac{8t}{3v_1 - 1} - \frac{3}{v_1^2}\right)(v_3 - v_1) \quad, \tag{10.7}$$

where we could have written v_3 instead of v_1 in the denominators. For the right-hand side of (10.6), by applying (10.5) we obtain

$$\int_{v_1}^{v_3} p(v)dv = \int_{v_1}^{v_3}\left(\frac{8t}{3v-1} - \frac{3}{v^2}\right)dv = \frac{8t}{3}\ln\frac{3v_3-1}{3v_1-1} + \frac{3}{v_3} - \frac{3}{v_1} \quad. \tag{10.8}$$

Equating the two sides gives

$$8t\left(\frac{v_3 - v_1}{3v_1 - 1} - \frac{1}{3}\ln\frac{3v_3 - 1}{3v_1 - 1}\right) - 3\frac{(v_3 - v_1)^2}{v_1^2 v_3} = 0 \quad. \tag{10.9}$$

Equation (10.9) contains both v_1 and v_3. These two quantities, however, are related by (10.5), since both must give the same pressure p_M:

$$\frac{8t}{3v_1 - 1} - \frac{3}{v_1^2} = \frac{8t}{3v_3 - 1} - \frac{3}{v_3^2} \quad,$$

$$\frac{24t(v_3 - v_1)}{(3v_1 - 1)(3v_3 - 1)} = \frac{3(v_3^2 - v_1^2)}{v_1^2 v_3^2} \quad,$$

$$\frac{8t}{(3v_1 - 1)(3v_3 - 1)} = \frac{v_3 + v_1}{v_1^2 v_3^2} \quad. \tag{10.10}$$

105

Solving (10.10) for v_3 and remembering that v_3 is the larger of the two solutions, one obtains:

$$v_3 = \frac{(3v_1 - 1)^2 + \sqrt{(3v_1 - 1)^4 - 4v_1(3v_1 - 1)(8tv_1^2 - 3(3v_1 - 1))}}{2(8tv_1^2 - 3(3v_1 - 1))} . \quad (10.11)$$

Using this equation we can replace v_3 in (10.9) by an expression in v_1 and t. The desired solution for v_1 is then obtained as the vanishing point of the function on the left-hand side of (10.9), at the specified temperature t. We cannot calculate this zero analytically, but we can find it on the computer using nested intervals.

For this method we need a lower and an upper limit for v_1. As the lower limit we take the reduced volume v_A corresponding to the point A in Fig. 10.2. We obtain v_A as the lowest zero of the discriminant occurring in (10.11), because the discriminant states whether there are any volumes other than v_1 giving the same pressure on the isotherm. The upper limit $v_{\lim}$ is taken as the reduced volume v_B corresponding to the zero point B in Fig. 10.2, whenever this zero point exists. It is easy to show that the zero point exists if the reduced temperature is lower than $\frac{27}{32}$. From the van der Waals equation (10.5) one obtains for $p = 0$

$$\frac{3}{v^2}(3v - 1) = 8t \quad , \quad 8tv^2 - 9v + 3 = 0 \quad , \quad v = \frac{9 \pm \sqrt{81 - 96t}}{16t} . \quad (10.12)$$

When $t \leq \frac{27}{32}$ there are two real solutions. The smaller one is the desired limit v_B. For reduced temperatures $t > \frac{27}{32}$ the isotherms have no zeros. In this case it can be shown analytically (or demonstrated numerically) that $v_{\lim} = 1$ fulfils all the conditions which we place on an upper limit, namely: a) $v_{\lim} = 1$ lies between the minimum and the maximum of the van der Waals isotherm, i.e. it is certainly greater than the left-hand end of the Maxwell line, b) equation (10.11) is valid in the entire interval $[v_A, 1]$ and fulfils the condition $v_3 > v_1$. As the upper limit $v_{\lim}$ we accordingly have $v_{\lim} = v_B$ if $t \leq \frac{27}{32}$ and $v_{\lim} = 1$ if $t > \frac{27}{32}$.

10.3 Programming

The method of nested intervals, which we shall use to determine the zero point of (10.9) taking account of (10.11), is also known by the name of the bisection method. It is applicable in the following problem formulation: given a continuous function $f(x)$ on an interval $[a, b]$, it is required to find a zero point x_0 of the function $f(x)$ with $a < x_0 < b$. The bisection method assumes that either $f(a) > 0$, $f(b) < 0$ or $f(a) < 0$, $f(b) > 0$, i.e. that at least one zero point of odd order exists. This assumption can be expressed in compact form by

$$f(a) \cdot f(b) < 0 \quad (10.13)$$

With $x_1 = a$ and $x_2 = b$ we come to the first step of the method: bisecting the interval $[x_1, x_2]$, one obtains

$$x_0 = \tfrac{1}{2}(x_1 + x_2) \tag{10.14}$$

and calculates $f(x_0)$.

The second step is to test whether, for a pre-specified accuracy limit ε,

$$|x_1 - x_2| \leq \varepsilon \ . \tag{10.15}$$

If this is the case, then x_0 is the desired approximation to the zero point. If this is not so, one tests whether the condition

$$f(x_0) \cdot f(x_2) < 0 \tag{10.16}$$

is fulfilled. If so, then a zero point lies between x_0 and x_2. We take $[x_0, x_2]$ as the new interval $[x_1, x_2]$ and return to the first step. If not, then either

$$f(x_0) = 0 \ , \tag{10.17a}$$

in which case we have found the required zero point, or else

$$f(x_1) \cdot f(x_0) < 0 \ . \tag{10.17b}$$

In the latter case we take $[x_1, x_0]$ as the new interval $[x_1, x_2]$ and return to the first step.

In order to be independent of our particular example, we do not incorporate the zero point determination into the main program KAP10, but write a subroutine with the name BISEKT.

Mathematical title	FORTRAN title	Mathematical title	FORTRAN title
f	F	x_1	X1
a	A	x_2	X2
b	B	$f(x_0)$	F0
ε	EPS	$f(x_1)$	F1
x_0	X0	$f(x_2)$	F2

Fig. 10.3. Notation employed in subroutine BISEKT

The subroutine BISEKT is shown in Fig. 10.4. It needs 4 input parameters, namely the function f, the interval limits a and b, and also the accuracy limit ε. As output one obtains (at most) one zero point x_0 (see list of parameters in line 100). The function f must be available as a FUNCTION subprogram $F(X)$. With the help of Fig. 10.3 one can follow the execution of the bisection method line by line. The progress of the calculation is guided by the four questions: (10.13) in line 106 (in the case of non-fulfilment the program stops), (10.15) in line 111, (10.16) in line 112 and (10.17) in line 116.

The function $f(v_1)$, whose zero point we wish to determine, stands on the left-hand side of (10.9),

```
        SUBROUTINE BISEKT(F,A,B,EPS,X0)                              100

        IMPLICIT DOUBLE PRECISION (A-H,O-Z)                          101
        X1=A                                                         102
        X2=B                                                         103
        F1=F(X1)                                                     104
        F2=F(X2)                                                     105
        IF(F1*F2.GE.0.D0)                                            106
     &  STOP 'Error in BISEKT: zero point cannot be found'           107
10      CONTINUE                                                     108
        X0=(X1+X2)/2.D0                                              109
        F0=F(X0)                                                     110
        IF(X2-X1.GT.EPS) THEN                                        111
          IF(F0*F2.LT.0.D0) THEN                                     112
            X1=X0                                                    113
            F1=F0                                                    114
            GOTO 10                                                  115
          ELSEIF(F0*F1.LT.0.D0) THEN                                 116
            X2=X0                                                    117
            F2=F0                                                    118
            GOTO 10                                                  119
          ENDIF                                                      120
        ENDIF                                                        121
        END                                                          122
```

Fig. 10.4. Subroutine BISEKT

$$f(v_1) = 8t\left(\frac{v_3 - v_1}{3v_1 - 1} - \frac{1}{3}\ln\frac{3v_3 - 1}{3v_1 - 1}\right) - 3\frac{(v_3 - v_1)^2}{v_1^2 v_3} \quad , \tag{10.18}$$

where v_3 is to be substituted using (10.11). The function $f(v_1)$ is defined in the interval $[v_A, v_{\lim}]$. In the bisection method one may extend the region of definition of the function f by appending functions, as long as no new zero points are created thereby. We make use of this possibility and extend the region of definition of $f(v_1)$ over the interval $[1/3,1]$, by setting

$$f(v_1) = +1 \quad \text{in the interval} \quad [1/3, v_A] \quad , \tag{10.19}$$

$$f(v_1) = -1 \quad \text{in the interval} \quad [v_{\lim}, 1] \quad . \tag{10.20}$$

The limit $v_{\lim}$ can easily be determined. From (10.20) one then has at once the function values in the (possibly null) interval $[v_{\lim}, 1]$.

The limit v_A, which would be more difficult to determine, need not be explicitly calculated. If, for an argument v_1 in the interval $[1/3, v_{\lim}]$, the discriminant of equation (10.11)

$$(3v_1 - 1)^4 - 4v_1(3v_1 - 1)(8tv_1^2 - 3(3v_1 - 1)) \quad , \tag{10.21}$$

has a negative value, then (10.19) is valid. If the value of the discriminant is greater than or equal to zero, then one calculates v_3 from (10.11) and $f(v_1)$ from (10.18). A typical profile of the function $f(v_1)$ is shown in Fig. 10.5.

We generate the function $f(v_1)$ by the FUNCTION sub-program FKT(V1) (Fig. 10.7, notation in Fig. 10.6). We transfer the temperature value needed for

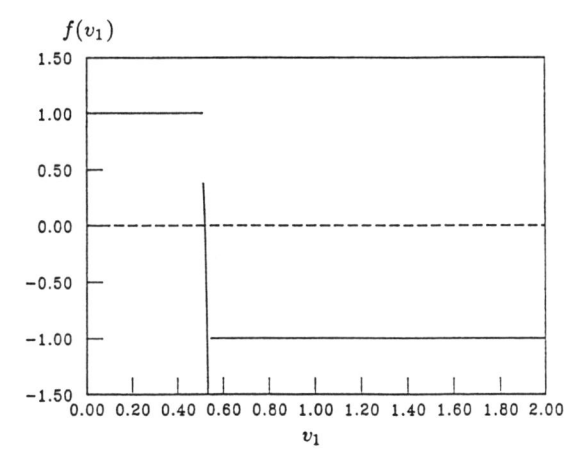

Fig. 10.5. Behaviour of the function $f(v_1)$ for the temperature $t = 0.8$

Physical title	FORTRAN title	Physical title	FORTRAN title
t	T	v_1	V1
$1/3$	B	v_3	V3
$v_{\lim}$	VLIM		

Fig. 10.6. Notation employed in FUNCTION subprogram FKT

```
      DOUBLE PRECISION FUNCTION FKT(V1)                       200
      IMPLICIT DOUBLE PRECISION (A-H,O-Z)                     201
      PARAMETER (B=1.D0/3.D0)                                 202
      COMMON /TEMPER/ T                                       203
      IF(T.GE.27.D0/32.D0) THEN                               204
        VLIM=1.D0                                             205
      ELSE                                                    206
        VLIM=(9.D0-SQRT(81.D0-96.D0*T))/(16.D0*T)             207
      ENDIF                                                   208
      IF(V1.LE.B) THEN                                        209
        FKT=1.D0                                              210
      ELSEIF(V1.GE.VLIM) THEN                                 211
        FKT=-1.D0                                             212
      ELSE                                                    213
        TERM=3.D0*V1-1.D0                                     214
        DENOM=8.D0*T*V1**2-3.D0*TERM                          215
        DISKRI=TERM**4-4.D0*V1*TERM*DENOM                     216
        IF(DISKRI.GE.0.D0) THEN                               217
          V3=(TERM**2+SQRT(DISKRI))/(2.D0*DENOM)              218
          FKT=8.D0*T*((V3-V1)/TERM-LOG((3.D0*V3-1.D0)/TERM)/3.D0)  219
     &         -3.D0*((V3-V1)/V1)**2/V3                       220
        ELSE                                                  221
          FKT=1.D0                                            222
        ENDIF                                                 223
      ENDIF                                                   224
      END                                                     225
```

Fig. 10.7. FUNCTION subprogram FKT

the computation by means of the COMMON block /TEMPER/ (line 203), as the calling program BISEKT does not know this variable. In order to economise in writing and computing we work with the following abbreviations: the variable TERM contains the value of the expression $3v_1 - 1$ (line 214), the variables DENOM and DISKRI relate to the denominator and the discriminant in equation (10.11) (lines 215 and 216).

One recognises in FKT the following equations: (10.11) in line 218, (10.12) in line 207, (10.20) in line 212 and (10.19) in line 222. The lines (209–212) ensure that FKT is defined for all values of the argument. One therefore avoids the question whether the argument lies in the region where the function is defined.

So far the discussion has been almost entirely about how the volumes v_1 and v_3 for the Maxwell line could be calculated. We shall now also consider the van der Waals isotherm in a definite volume interval $[v_{\min}, v_{\max}]$. For this purpose we shall divide this interval into $i_v = 200$ sub-intervals and obtain $i_v + 1$ grid points.

$$v^{(i)} = ih + v_{\min} \quad , \quad i = 0, 1, 2, \ldots, i_v \quad , \tag{10.22}$$

with the mesh width

$$h = \frac{v_{\max} - v_{\min}}{i_v} \quad . \tag{10.23}$$

We notice, meanwhile, that $v_{\min}$ must be greater than $\frac{1}{3}$. The values of the pressures at the grid points follow from (10.5) and are

$$p^{(i)} = \frac{8t}{3v^{(i)} - 1} - \frac{3}{(v^{(i)})^2} \quad . \tag{10.24}$$

The pressure p_M at the Maxwell line is obtained in the same way:

$$p_M = \frac{8t}{3v_1 - 1} - \frac{3}{v_1^2} \quad . \tag{10.25}$$

We now consider the main program KAP10 (Fig. 10.9). The values for i_v and ε, together with the eigenvolume $\frac{1}{3}$ are specified in a PARAMETER instruction (line 302); the values for $v_{\min}$, $v_{\max}$ and t are input.

Physical title	FORTRAN title	Physical title	FORTRAN title
1/3	B	ε	EPS
t	T	p_M	PM
$v_{\min}$	VMIN	i_v	IV
$v_{\max}$	VMAX	h	H
v_1	V1	$v^{(i)}$	V(I)
v_3	V3	$p^{(i)}$	P(I)

Fig. 10.8. Notation employed in main program KAP10

```
      PROGRAM KAP10                                              300

      IMPLICIT DOUBLE PRECISION (A-H,O-Z)                        301
      PARAMETER (IV=200, B=1.D0/3.D0, EPS=1.D-14)                302
      DIMENSION V(0:IV), P(0:IV)                                 303
      COMMON /TEMPER/ T                                          304
      EXTERNAL FKT                                               305
*     (Input: VMIN,VMAX)

      H=(VMAX-VMIN)/IV                                           306
      DO 10 I=0,IV                                               307
         V(I)=I*H+VMIN                                           308
10    CONTINUE                                                   309

*     (Input: T)

      DO 20 I=0,IV                                               310
         P(I)=8.D0*T/(3.D0*V(I)-1.D0)-3.D0/V(I)**2               311
20    CONTINUE                                                   312

*     (Output: V,P)

      IF(T.LT.1.D0) THEN                                         313
         CALL BISEKT(FKT,B,1.D0,EPS,V1)                          314
         TERM=3.D0*V1-1.D0                                       315
         DENOM=8.D0*T*V1**2-3.D0*TERM                            316
         DISKRI=TERM**4-4.D0*V1*TERM*DENOM                       317
         V3=(TERM**2+SQRT(DISKRI))/(2.D0*DENOM)                  318
         PM=8.D0*T/(3.D0*V1-1.D0)-3.D0/V1**2                     319

*     (Output: V1,V3,PM)

      ENDIF                                                      320
      END                                                        321
```

Fig. 10.9. Numerical portion of the main program KAP10

After the grid points have been calculated from (10.22) in the DO-loop of lines 307 to 309, the pressure values on the van der Waals isotherm are calculated from (10.24) in the DO-loop of lines 310 to 312. If the temperature is smaller than the critical temperature, i.e. $t < 1$, the subroutine BISEKT is called (line 314), which transfers the value of v_1 to KAP10. With the abbreviations TERM, DENOM and DISKRI already known, v_3 is calculated from (10.11) in line 318, and in the following line we obtain p_M from (10.25).

We wish to output the isotherms graphically. We do this, as before, using GMOVE and GDRAW. The isotherm $t = 1$ is automatically calculated and output as the first curve, provided that it intersects the graphics window. Subsequently the user can request further isotherms. In addition to the isotherm we output the reduced temperature t of the isotherm most recently drawn and the vapour pressure p_M, using GWRITE. In contrast to Chap. 7, we wish the two quantities to appear, not in the upper margin of the screen, but in the graphics window. We therefore call GWRITE in the following way:

CALL GWRITE(2,X,Y,NOTE,ICOLOR) .

When the first parameter of GWRITE is equal to 2, the coordinates X and Y do not refer to the whole screen, but to the graphics window. The points (0,0) and (100,100) thus correspond to the bottom left, and the top right-hand corners, respectively, of the graphics window. The meaning of the other input parameters is the same is in Chap. 7.

We introduce a new method for the input of the temperature t. Instead of inputting t directly, we input a pair of values (p, v), and the program calculates the associated temperature value from the van der Waals equation (10.15). We input the pair of values (p, v) graphically by means of a graphic cursor (small cross wires). For this we do not have to leave the graphics mode, so that all the curves already drawn are retained.

This form of input is made available by the subroutine GINPUT. It is activated by the call

CALL GINPUT(NOTE, XSTART, YSTART, X, Y, INPUT)   .

As already mentioned in Chap. 4, NOTE is the CHARACTER expression for a text which is displayed above the graphics. We use the coordinates XSTART and YSTART to determine the start position of the cursor inside the diagram area. By means of the cursor keys (arrow keys) we move the cursor to the desired position. The cursor can usually be moved over the screen in either large or small steps according to choice; most personal computers use the enter key to switch between the two possibilities. If the enter key or a letter, numeral or symbol key is pressed, GINPUT gives control back to the calling program, transferring the current cursor coordinates into the variables X and Y. The parameter INPUT contains the ASCII value of the key pressed.

In the program KAP10 we use the parameter INPUT in the following way: depressing the enter key (ASCII value 13) signifies "Calculate and draw selected isotherm", depressing "Q" or "q" (ASCII values 81 and 113, respectively) signifies "Terminate program". Other keys are ignored, i.e. GINPUT is again called.

The complete program briefly summarised here is to be found on the diskette in the file KAP-10.FOR.

10.4 Exercises

10.4.1 Consider the p-v diagram for a number of different temperatures. Find the vapour pressure p_M for various reduced temperatures t. Draw the function $p_M(t)$ between $t = 0.25$ and $t = 1$ on a logarithmic scale.

10.4.2 Carbon dioxide has a critical temperature of $31.1°$ C and a critical pressure of 72.95 atm. (see [10.1]). Calculate the critical volume according to van der Waals. Compare the volume of one mole of CO_2 according to the ideal gas equation and according to van der Waals, at $40°$ C and for 1, 10, 100 and 1000 atmospheres.

Note: One mole of ideal gas at one atmosphere and $0°$ C has a volume of $22420 \, \text{cm}^3$, and consequently at one atmosphere and $40°$ C it has a volume of $25700 \, \text{cm}^3$.

10.4.3 Figure 10.10 shows the measured boiling temperatures for the gases CH_4, NH_3, H_2O, HF and Ne at various pressures. Find the boiling temperatures according to van der Waals, using the curve obtained in Exercise 10.4.1. For which gases do you find good agreement, and for which poor?

Gas	CH$_4$	NH$_3$	H$_2$O	HF	Ne
Molecular weight	16.04	17.03	18.02	20.01	20.18
Crit. pressure (atm)	47.2	115.2	218.0	66.2	27.1
Crit. temperature (K)	190.6	405.5	647.4	461	44.4
Boiling temperature (K) measured					
at 1 atm	111.7	239.7	373.2	292.7	27.07
at 2 atm		254.5	393.3	313.3	29.63
at 5 atm	134.9	277.9	425.6	343.1	33.81
at 10 atm	148.4	298.9	453.7	371.5	37.65
at 20 atm	164.7	323.3	486.3		42.35
at 40 atm	186.9	352.1	524.3		

Fig. 10.10. Measured boiling temperatures of CH$_4$, NH$_3$, H$_2$O, HF, Ne (from [10.2])

10.4.4 Propane (C$_3$H$_8$), butane (C$_4$H$_{10}$) and mixtures of the two are handled as liquid gas in canisters for heating of camp-cookers, soldering-irons, etc. Notices on the canisters warn the user not to subject the containers to any high temperatures. Using the vapour pressure curve from Exercise 10.4.2, estimate to what extent these warnings are justified. How high is the pressure according to van der Waals at 20° C, and how much does it increase if the container is heated to 50, 70 or 100° C?

Gas	C$_3$H$_8$	C$_4$H$_{10}$
Molecular weight	44.10	58.12
Crit. pressure (atm)	42.1	37.4
Crit. temperature (°C)	96.8	152.01

Fig. 10.11. Critical data for liquid gases in household use (from [10.2])

10.5 Solutions to the Exercises

10.5.1 Figure 10.12 shows the curve of $p_M(t)$. Since $p_M(t)$ varies by about three orders of magnitude between $t = 0.25$ and $t = 1$, a logarithmic scale is required. As one can see, in the neighbourhood of the critical point a fall of about 15 % in t leads to a halving of the vapour

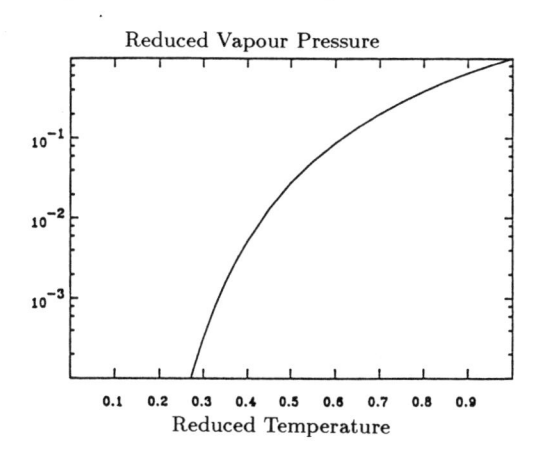

Fig. 10.12. Vapour pressure curve $p_M(t)$

pressure. Further below the critical point a fall of a few per cent in t is enough to halve the vapour pressure.

10.5.2 The exercise can be carried out by the program most simply if the desired pressure is chosen as the lower or upper limit of the pressure region displayed in the graphics window. After some calculation one obtains the following values for the volume of a mole at 40° C:

Volume of a mole in cm^3

	ideal gas equation	VAN DER WAALS equation	measured (from [10.1])
at 1 atm	25700	25590	25574
at 10 atm	2570	2470	2448
at 100 atm	257	89	69.3
at 1000 atm	25.7	54	40

We see that especially at higher pressures the volumes of a mole determined from the van der Waals equation lie nearer to the experimental values than those determined from the ideal gas equation. That the agreement cannot be perfect is clear from the fact that the values $P_c = 72.95$ atm and $T_c = 31.1°$ C lead to a critical volume for the mole $V_c = 128$ cm^3, whereas the experimentally determined value for V_c is about 100 cm^3.

10.5.3 The calculated boiling temperatures do not agree particularly well with the experimental values for any of the gases. The agreement is particularly poor for H_2O, nearly as bad are those of NH_3 and HF, and hence for the molecules with a permanent electric dipole moment. Accurate agreement is indeed not to be expected, however, since the van der Waals equation was developed as an improvement for the ideal gas equation, seeking to describe better the behaviour of real gases by the introduction of only two new parameters.

Gas	CH$_4$	NH$_3$	H$_2$O	HF	Ne
Boiling temperature (K) calculated					
at 1 atm	92	173	240	212	23
at 2 atm	102	190	261	233	26
at 5 atm	118	217	295	270	31
at 10 atm	135	243	328	306	35
at 20 atm	156	276	368	351	41
at 40 atm	183	318	418	408	

Fig. 10.13. Calculated boiling temperatures for CH_4, NH_3, H_2O, HF, Ne

10.5.4 We obtain the following vapour pressures:

Temperature (°C)	0	20	50	70	100
Vapour pressure (atm)					
for propane (C_3H_8)	11.0	15.4	23.9	30.9	
for butane (C_4H_{10})	4.8	6.9	11.3	14.9	21.7

A rise from 20° C to 70° C more than doubles the pressure in the container for both gases. The actual increase is even greater than that calculated from the van der Waals equation. Since even 70° C can be exceeded if one leaves the containers in the boot of a car in the sunshine, the warnings are thoroughly justified.

11. Solution of the Fourier Heat Conduction Equation and the "Geo-Power Station"

11.1 Formulation of the Problem

The Fourier heat conduction equation describes the variation of temperature with time in a rigid body. It is of historical interest. It appeared for the first time in 1822 in a paper presented to the Académie Française by Fourier entitled "Théorie de la Chaleur". The work was considered at that time to be very important, because complete equilibrium of temperature in the universe was regarded as the heat-death of the world, i.e. as a possible form of universal decline. Today Fourier's work is important for another reason. In fact, as a method of solution for the heat conduction equation, Fourier set out in his paper the theory of Fourier Series.

We shall solve the heat conduction equation using the method proposed by Fourier, and shall first look for a suitable example. The heat-death of the universe has become less interesting in view of other possible forms of world decline. Much more topical is the question of the energy supply for mankind. As an alternative energy source it has been often suggested that one should simply exploit the heat from the hot interior of the earth. That this is possible in principle is demonstrated by the many natural hot springs. Where geological anomalies are present, as in Tuscany, for example, artificial hot springs can be successfully used as energy sources. The question arises whether one could manage even without geological anomalies, i.e. whether it is possible to exploit the heat from the dry hot rock in deep strata.

To test the technical feasibility of abstracting heat from deep rock strata, the "Hot Dry Rock Process" has been devised in the USA. A borehole is sunk several thousand metres deep and by injecting water under high pressure one produces deep down a layer of rock with multiple fractures, called "Fracs". The Fracs are pumped out through another borehole in a different place. By maintaining the high water pressure the Frac can now be used as a boiler. Cold water is injected through the first borehole, hot water is extracted from the second borehole. The heat produced is fed, either directly or via a heat exchanger, to a power station.

By means of a computer model we shall try out the Hot Dry Rock Process, and pose the question, how large an engineering effort is required in order to extract as much heat as is required by a large modern power station (1300 Megawatt electric power).

A rule of thumb in geology states that the temperature of the earth increases with increasing depth by about 3° C per 100 m. In our model we can accordingly bore to 6000 m, 8000 m or 10 000 m deep and — with a temperature of 10° C at the surface — obtain a temperature at depth of 190° C, 250° C or 310° C. In order to simplify the calculation, we shall give the fractures (Fracs) in the model a simple shape and arrangement. They will be of uniform shape and arranged periodically. Each Frac will be parallel to, and at a distance d from, the next Frac. More precisely, between the parallel faces of each pair of Fracs there is a layer of rock of thickness d, the heat from which is carried away by water flowing through the Fracs. The temperature profile at time $t = 0$ is assumed to be as shown in Fig. 11.1. The rock has the temperature T_0, the water the temperature $T_1 < T_0$. The thickness d of the rock layer, the temperature difference $T_0 - T_1$ and also the total surface area of the rock/water interface are input parameters of our program. When the power station is in operation the temperature T_1 is kept constant as the working temperature of the power station. One can achieve this in the model by allowing very fast flow of water through the Frac at the start, and decreasing the water flow as the rock cools. The result is that the output of the power station varies with time. We shall calculate this hypothetical decrease. It is left to the reader to make the program more realistic by first choosing a working temperature lying close to T_0, and then lowering this with the passage of time. By this means the power station can be made to operate at constant power; the singularities in the output caused by abrupt changes in the working temperature are harmless and can be smoothed out either by hand or by computer.

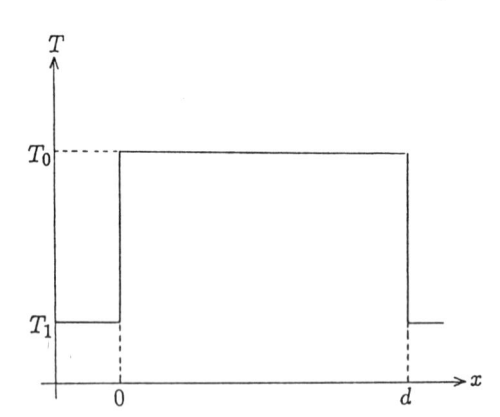

Fig. 11.1. Temperature profile at time $t = 0$

The temperature of the rock varies with position and time as a result of the heat extraction. We shall calculate the temperature profile $T(x, t)$, and also the thermal energy $P(t)$ transferred each hour from the rock to the water.

First of all we need the heat transfer equation. It can easily be derived. One assumes a specified temperature distribution $T(\mathbf{r}, 0)$ in a rigid body at time $t = 0$, and wishes to find the temperature distribution $T(\mathbf{r}, t)$ at some later time t. For the flow of heat Fourier postulated the expression

$$q(r,t) = -\lambda \operatorname{grad} T(r,t) \tag{11.1}$$

with λ [W/(mK)] denoting the heat conduction coefficient. The heat flow is proportional to the temperature gradient and in the reverse direction. The heat conduction coefficient specifies the quantity of heat flowing perpendicularly across a surface of $1\,\text{m}^2$ in unit time as a result of a temperature gradient of $1\,\text{K/m}$. By taking the divergence, one obtains from (11.1)

$$\operatorname{div} q(r,t) = -\lambda \Delta T(r,t) \ . \tag{11.2}$$

The divergence of a flow of heat describes the outflow of heat from a unit volume. Positive divergence indicates cooling. The cooling per unit time is smaller, the greater the specific heat c and the greater the mass per unit volume, i.e. the greater the density ϱ of the body. Fourier set

$$\operatorname{div} q(r,t) = -\varrho c \frac{\partial T}{\partial t}(r,t) \tag{11.3}$$

and hence obtained the heat conduction equation

$$\frac{\partial T}{\partial t}(r,t) = \frac{\lambda}{\varrho c} \Delta T(r,t) \ . \tag{11.4}$$

In our computer model this equation is simplified because of the idealised arrangement of the Fracs. The heat can only flow perpendicularly to the rock/water interface. We set the x-axis in this direction and obtain the differential equation for one-dimensional heat conduction

$$\frac{\partial T}{\partial t}(x,t) = \frac{\lambda}{\varrho c} \frac{\partial^2 T}{\partial x^2}(x,t) \ . \tag{11.5}$$

11.2 Method of Solution

We seek a solution $T(x,t)$ of (11.5) under the following boundary conditions:

1. The temperature distribution $T(x,0)$ is specified at $t=0$.
2. $T(0,t) = T(d,t) = T_1$ when $t > 0$.

Particular solutions $T^{(n)}$ of (11.5), which assume values independent of time at $x=0$ and $x=d$, are easily found. By substitution of

$$T^{(n)}(x,t) = \sin\left(\frac{n\pi}{d}x\right) a_n(t) + \text{const.} \quad , \quad n = 1, 2, \ldots \quad , \tag{11.6}$$

in (11.5) we obtain

$$\frac{da_n(t)}{dt} = -\frac{\lambda}{\varrho c}\left(\frac{n\pi}{d}\right)^2 a_n(t) \ . \tag{11.7}$$

The solution of this differential equation is

$$a_n(t) = a_n(0) \exp\left[-\frac{\lambda}{\varrho c}\left(\frac{n\pi}{d}\right)^2 t\right] \quad . \tag{11.8}$$

With (11.8), (11.6) becomes a particular solution of (11.5); the constant is determined by the second of the two boundary conditions.

Fourier's method of solution consists of superposing particular solutions so that the specified temperature distribution $T(x,0)$ is reproduced at the starting time $t = 0$. Since any linear combination of particular solutions is itself a solution of the equation, one thus obtains a soluton of (11.5) which also fulfils the first of the two boundary conditions.

We form the superposition

$$T(x,t) = T_1 + \sum_{n=1}^{\infty} a_n(t) \sin\left(\frac{n\pi}{d} x\right) \quad . \tag{11.9}$$

The coefficients $a_n(0)$ are fitted to the initial temperature distribution $T(x,0)$.

The temperature distribution $T(x,0)$ is assumed to take the rectangular form shown in Fig. 11.1,

$$\begin{aligned} T(x,0) &= T_1 \quad , \quad \text{for} \quad x=0 \quad \text{and} \quad x=d \\ T(x,0) &= T_0 \quad , \quad \text{for} \quad 0 < x < d \quad . \end{aligned} \tag{11.10}$$

Because of the infinite temperature gradient at $x = 0$ and $x = d$, this temperature distribution does not exactly represent physical reality, but this will have no appreciable effect on the results of our calculation. If we were actually to take an infinite series in (11.9), then the solution function $T(x,t)$ would reproduce the abrupt rise at $x = 0$ and $x = d$ for $t = 0$, and the heat flow would be infinitely great when $t = 0$. For the numerical treatment of the problem, however, we have to curtail the infinite series after a certain number $n_{\max}$ of terms. This has the result that when $t = 0$ the rectangular distribution is reproduced only approximately, i.e. with slight waves and with rounded corners. The problem of the infinitely great flow of heat when the power station starts up is thereby avoided. The curtailment of the infinite series does indeed introduce an error, but this can be estimated. The expansion coefficients $a_n(t)$ with $n > n_{\max}$ decrease very rapidly with increasing time, as one can see from (11.8); only directly after the start up of the power station do they have an appreciable effect on the solution function $T(x,t)$. If we choose $n_{\max}$ large enough, the error in the temperature distribution becomes negligible very soon after start up.

The advantage of the rectangular distribution (11.10) is that for this distribution the integrations in the calculation of the Fourier coefficients $a_n(0)$ are particularly easy to carry out.

$$\begin{aligned} a_n(0) &= \frac{2}{d} \int_0^d dx (T(x,0) - T_1) \sin\left(\frac{n\pi}{d} x\right) = \frac{2}{d} \int_0^d dx (T_0 - T_1) \sin\left(\frac{n\pi}{d} x\right) \\ &= \begin{cases} \dfrac{4}{n\pi}(T_0 - T_1) & \text{for} \quad n = 1, 3, 5, \ldots , \\ 0 & \text{for} \quad n = 2, 4, 6, \ldots \end{cases} \end{aligned} \tag{11.11}$$

We can now substitute (11.8) into (11.9) and obtain as the solution

$$T(x,t) = T_1 + \sum_{n=1}^{\infty} a_n(0) \sin(k_n x) \exp(-b_n t) \quad, \tag{11.12}$$

with the coefficients $a_n(0)$ from (11.11) and the abbreviations

$$k_n = \frac{n\pi}{d} \quad \text{and} \tag{11.13}$$

$$b_n = \frac{\lambda}{\varrho c}\left(\frac{n\pi}{d}\right)^2 \quad. \tag{11.14}$$

11.3 Programming

In order to calculate the temperature profile $T(x,t)$, we divide the interval $[0,d]$ of the x-axis into i_d sub-intervals and in this way obtain $i_d + 1$ mesh points,

$$x_i = ih \quad, \quad i = 0, 1, 2, \ldots, i_d \quad, \tag{11.15}$$

with the mesh width

$$h = \frac{d}{i_d} \quad. \tag{11.16}$$

The temperature values at these mesh points are given from (11.12) as

$$T_i(t) = T(x_i, t) = T_1 + \sum_{n=1}^{n_{\max}} a_n(0) \sin(k_n x_i) \exp(-b_n t) \quad. \tag{11.17}$$

In Fig. 11.3 we see how the calculation of the temperature values $T_i(t)$ is carried out in the main program KAP11; the notation employed is to be found in Fig. 11.2. We choose i_d to be 100 (line 103). In line 107 the mesh width

Physical title	FORTRAN title	Physical title	FORTRAN title
λ	LAMBDA	$a_n(0)$	ANO
ϱ	RHO	b_n	BN
c	C	k_n	KN
T_0	T0	$\exp(-b_n t)$	EN
T_1	T1	i_d	ID
d	D	h	H
A	A	x_i	X(I)
t	TIME	T_i	T(I)
n	N	P	P
$n_{\max}$	NMAX	Q	Q
		π	PI

Fig. 11.2. Notation employed in main program KAP11

```
            PROGRAM KAP11                                         100

            IMPLICIT DOUBLE PRECISION (A-H,O-Z)                   101
            DOUBLE PRECISION LAMBDA, KN                           102
            PARAMETER (ID=100, PI=3.1415926535D0)                 103
            PARAMETER (LAMBDA=2.5D-6, RHO=2500.D0, C=2.35D-7)     104
            PARAMETER (EXPMAX=50.D0)                              105
            DIMENSION X(0:ID), T(0:ID)                            106

*           (Input: T0, T1, D, A, TIME, NMAX)

            H=D/ID                                                107
            DO 10 I=0,ID                                          108
             X(I)=I*H                                             109
             T(I)=T1                                              110
      10    CONTINUE                                              111
            P=0.D0                                                112
            Q=0.D0                                                113
            DO 30 N=1,NMAX,2                                      114
             ANO=4.D0/(N*PI)*(T0-T1)                              115
             KN=N*PI/D                                            116
             BN=LAMBDA*KN*KN/(RHO*C)                              117
             IF(BN*TIME.LE.EXPMAX) THEN                           118
              EN=EXP(-BN*TIME)                                    119
             ELSE                                                 120
              EN=0.D0                                             121
             ENDIF                                                122
             DO 20 I=0,ID                                         123
              T(I)=T(I)+ANO*SIN(KN*X(I))*EN                       124
      20     CONTINUE                                             125
             P=P+ANO*KN*EN                                        126
             Q=Q+ANO*KN/BN*(1.D0-EN)                              127
      30    CONTINUE                                              128
            P=LAMBDA*A*P                                          129
            Q=LAMBDA*A*Q                                          130

*           (Output: X, T, P, Q)

            END                                                   131
```

Fig. 11.3. Numerical portion of main program KAP11

h is determined from (11.16). The mesh points are calculated from (11.15) in the DO-loop of lines 108 to 111. In this DO-loop also the temperatures $T_i(t)$ are initialised with the temperature T_1 (see (11.17)). The summation over n is effected in the DO-loop of lines 114 to 128. Since the coefficients $a_n(0)$ vanish for even n, we need only sum over odd values of n (see line 114). In lines 115 to 117 the coefficients $a_n(0)$, k_n and b_n are calculated. In the calculation of the exponential function in (11.17) exponent overflow can occur. Several compilers (e.g. the PROFESSIONAL FORTRAN compiler) provide a diagnostic message for this eventuality. By means of the IF-block of lines 118 to 122 we suppress the diagnostic message and set $\exp(-b_n t) = 0$ if the argument $b_n t$ exceeds a specified maximum value EXPMAX. The setting of EXPMAX in line 105 should be adjusted to the computer in use. By the specification of EXPMAX we also have the possibility of avoiding the computation of the exponential function when this would be too small to have any influence. Finally the temperature values $T_i(t)$ are calculated from (11.17) in the DO-loop of lines 123 to 125.

After we know the temperature profile at a specified time, we are naturally interested to know the heat flow $P(t)$ at this time, and the thermal energy $Q(t)$ so far produced. The area of heat exchange surface is denoted by A. From the thermal flux

$$|q(0,t)| = \frac{1}{A}\frac{dQ(t)}{dt} \tag{11.18}$$

we obtain, using (11.1) and (11.12), the relation

$$P(t) = \frac{dQ(t)}{dt} = A|q(0,t)| = A\lambda\frac{\partial T(0,t)}{\partial x}$$

$$= A\lambda \sum_{n=1}^{\infty} a_n(0) k_n \exp(-b_n t) \quad . \tag{11.19}$$

For the thermal energy $Q(t)$ so far produced we accordingly obtain

$$Q(t) = \int_0^t P(\tau)d\tau = A\lambda \sum_{n=1}^{\infty} \frac{a_n(0)k_n}{b_n}(1 - \exp(-b_n t)) \quad . \tag{11.20}$$

Equation (11.19) is evaluated in lines 112, 126 and 129, and (11.20) in lines 113, 127 and 130.

For granite the values of the thermal conduction coefficient λ, the density ϱ and the specific heat c are about

$$\lambda = 2.5 \times 10^{-6}\frac{\mathrm{MW}}{\mathrm{m\,K}} \quad , \quad \varrho = 2500\frac{\mathrm{kg}}{\mathrm{m^3}} \quad , \quad c = 2.35 \times 10^{-7}\frac{\mathrm{MWh}}{\mathrm{kg\,K}} \quad .$$

These values are specified in the program (line 104). The values for the temperatures T_0 and T_1, the rock thickness d, the surface area of the rock layer A, the time t and the summation index $n_{\max}$ are input. In the input of the time t one has the option of choosing between hours, days and years as the units of time. Immediately after the input of the time this is converted into hours, since the numerical portion of the program uses hours exclusively as units of time. The choice of the time units is achieved by means of a CHARACTER variable, with which we have become familiar in Chap. 2 in connection with a Yes/No decision.

The temperatures are output in graphical form and as an option subsequently in numerical form. The values for the heat flow and the thermal energy are included in the graphics. The complete program summarised here is to be found on the diskette in the file KAP-11.FOR.

11.4 Exercises

11.4.1 Test the accuracy of the Fourier method for the following studies:

a) How large must $n_{\max}$ be chosen for the specified temperature profile at time $t=0$ to be reproduced with sufficient accuracy?

b) How does the choice of the rock thickness d influence the accuracy of the solution?
c) How sensitively does the thermal output $Q(t)$ depend on variations in n_{max} and d?

11.4.2 Calculate the temperature profile and the heat flow for various temperatures T_0, T_1 and for various elapses of time t. How great is the output of the model power station after 2, 5, 10 years with a borehole depth of 10 000 m and a rock surface area of 30 km^2?

11.4.3 Discuss the differences between the computational model and a realistic geo-power station. Could a realistic power station work substantially more effectively than the model?

11.5 Solutions to the Exercises

11.5.1 a) We set $T_0 = 300°$ C, $T_1 = 200°$ C, $d = 50$ m and $t = 0$. With $n_{max} = 50$ we obtain the left one of the temperature curves shown in Fig. 11.4. Instead of a rectangle (see Fig. 11.1) we obtain a wavy line, i.e. we have curtailed the Fourier series (11.12) too early. With $n_{max} = 1000$ we obtain the right one of the temperature curves shown in Fig. 11.4. The accuracy of reproduction of the rectangle has become very much better. The waviness of the temperature curve is no longer detectable. The numerical output of the temperature curve, however, shows that the abrupt rise of temperature in the neighbourhood of $x = 0$ is even now not exactly reproduced.

b) As is apparent from (a), the distribution of initial temperature specified in Fig. 11.1 is replaced in the numerical calculation by an approximate distribution. How thick the layer of rock is, which is affected by the inaccuracy of the approximation, depends on the total thickness d. Whilst the inaccuracy has an extent of a few centimetres when $d = 50$ m, it becomes a few metres when $d = 5000$ m. One should therefore not choose d greater than necessary.

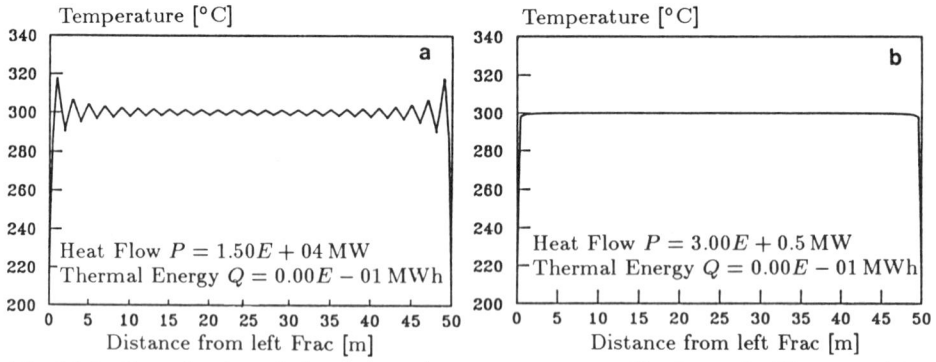

Fig. 11.4. Reproduction of the rectangular temperature profile shown in Fig. 11.1 with $n_{max} = 50$ (left picture) and $n_{max} = 1000$ (right picture)

c) The inexact reproduction of the rectangular temperature profile at $t = 0$ is caused by the curtailment of the Fourier series. The neglected terms of the series die away with time faster than the terms which have been included. The inaccuracy therefore makes the energy transfer behave as though the power station had been running for a short time before its start up. The energy transferred in that time is missing from $Q(t)$. Since a great deal of heat flows at the moment of start up, this leads to a sensitivity of $Q(t)$ to n_{max} and d. With a suitable choice of these latter quantities one will obtain reliable values for $Q(t)$. For longer operating times of the power station the first terms in the series (11.17) become dominant.

The heat extracted in the second and third years scarcely depend any more on n_{max}, unless one has input too large a value for d.

11.5.2 Corresponding to a borehole depth of 10 000 m we set $T_0 = 310°$ C. With $T_1 = 200°$ C, $d = 160$ m and $t = 5$ years we obtain the temperature profile shown in Fig. 11.5. A layer of rock of about 40 m thickness shows a marked change of temperature. Interior layers of rock have been less affected by the transfer of heat to the power station. After 2, 5, 10 years the thermal production of the power station amounts to 539, 341, 241 Megawatts. The considerable decrease in power over the course of a year results from the fact that the heat has to be transferred across ever thicker layers of rock in order to reach the cooling liquid. A thermal output of 241 MW is very low, when one thinks of the engineering effort required to produce a rock surface of 30 square kilometers (!) at great depth. In the conversion of the thermal energy into electric energy, the low efficiency associated with these meagre working temperatures would also play an important role.

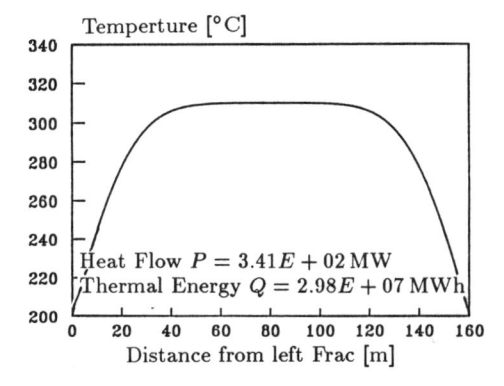

Fig. 11.5. Temperature profile after 5 years

Heat Flow $P = 3.41E + 02$ MW
Thermal Energy $Q = 2.98E + 07$ MWh

11.5.3 In a realistic geo-power station T_1 would not have a fixed value either spacewise or timewise. At the place where the water was injected T_1 would have a value of about 30° C. In this region the rock would be cooled more than in the example of Exercise 11.4.2. In the region of water extraction one would strive for the highest possible value for T_1, which would increase the efficiency of conversion of heat into electrical energy. Our model could from this point of view be improved. If one is interested only in the question, whether the energy won from a geo-power station of the type here described is worth the engineering expenditure, then it is a question of powers of ten, and a factor of 2 or 3 plays a minor role. The results of the simple model can certainly serve as the basis for more detailed discussions.

12. Group and Phase Velocity in the Example of Water Waves

12.1 Formulation of the Problem

As a preparation for quantum mechanics we shall in this chapter consider the propagation of waves. As the basic wave form we take a plane, scalar wave of infinite extent and sinusoidal profile. If we choose the x-axis in the direction of propagation, the wave function becomes

$$\psi(x,t) = \psi_0 \sin(kx - \omega t) \quad , \quad \text{or} \quad \psi(x,t) = \psi_0 \cos(kx - \omega t) \quad . \tag{12.1}$$

The modulus of the wave number k and the angular frequency ω are determined by the wavelength λ and the wave period τ:

$$|k| = \frac{2\pi}{\lambda} \quad , \quad \omega = \frac{2\pi}{\tau} \quad . \tag{12.2}$$

If one follows the maximum elevation of the wave with time, this moves with the velocity

$$v_p = \frac{\omega}{k} \quad . \tag{12.3}$$

The velocity v_p is called the phase velocity. In nature there are no infinite plane waves. Any wave motion occurring in nature has a beginning and an end, both with regard to space and time. One speaks in this case of a wave train or a wave packet. Following Fourier we can represent a wave packet as a superposition of plane waves. If we again consider only propagation in the x-direction and take a wave packet that is an even function of x at time $t = 0$, then we can write

$$\psi(x,t) = \int_{-\infty}^{\infty} a(k) \cos(kx - \omega t) dk \quad . \tag{12.4}$$

The function $a(k)$ determines the form of the wave packet and also decides in which direction it moves. In many fields of mechanics, acoustics, optics and quantum mechanics the principle of superposition of waves is valid: the linear superposition (12.4), as well as its individual components, satisfy the relevant wave equation. The superposition principle, however, is not valid for all wave motions. For example, it is not valid for water waves which have breaking crests.

The angular frequency ω is a function of the wave number k,

$$\omega = \omega(k) \quad . \tag{12.5}$$

The properties of this function are characteristic for the physical behaviour of the wave propagation.

The simplest case is the propagation of light in a vacuum. Light waves are not indeed scalar waves, but one can, for example, identify one component of the electric field strength with $\psi(x,t)$, and then one has for this component the wave functions presented in (12.1) and (12.4). In the propagation of light in a vacuum ω is a linear function of the modulus of the wave number,

$$\omega(k) = c|k| \,, \tag{12.6}$$

where c is the velocity of light. By substituting this relation in (12.4) one obtains

$$\psi(x,t) = \int_0^\infty a(k) \cos [k(x-ct)]dk + \int_{-\infty}^0 a(k) \cos [k(x+ct)]dk \,. \tag{12.7}$$

If we restrict ourselves to wave packets which propagate in the positive x-direction, the second of the two integrals is discarded. The first integral is unchanged if one increases the time t by Δt and at the same time moves along the x-axis a distance $c \cdot \Delta t$. The wave packet moves with the velocity of light and does not change its shape.

When light propagates through matter, the function $\omega(k)$ takes a more general form. With normal dispersion and long, almost monochromatic wave trains, one gets a good approximation if one expands $\omega(k)$ in a Taylor series about the mean wave number k_0 of the wave packet and curtails it after the linear term,

$$\omega(k) = \omega(k_0) + (k-k_0)\frac{d\omega}{dk}\bigg|_{k=k_0} + O((k-k_0)^2) \,. \tag{12.8}$$

If one still considers only wave packets propagating in the positive x-direction, substitution of this relation in (12.4) gives

$$\psi(x,t) = \int_0^\infty a(k) \cos [k(x-\omega'(k_0)t) - \{\omega(k_0)t - k_0\omega'(k_0)t\}]dk \,. \tag{12.9}$$

The expression in brace brackets is a phase shift, which depends on time, but not on the wave number k. The phase shift has little influence on the shape of the wave packet. We can, for example, consider the wave packet when the phase shift is just equal to an integral multiple of 2π. Then we see that the wave packet moves with the velocity

$$v_g = \frac{d\omega}{dk}\bigg|_{k=k_0} \equiv \omega'(k_0) \,. \tag{12.10}$$

The velocity v_g is called the group velocity, since the wave packet, or wave group, moves with this velocity. A Morse signal consists of a sequence of wave packets. Since they propagate with the velocity v_g one also calls v_g the signal velocity. With normal dispersion one always has

$$v_g \leq c \,. \tag{12.11}$$

In the motion of very short wave packets, i.e. of wave packets which consist of only a few wave crests and troughs, the terms $O((k - k_0)^2)$ in the Taylor series (12.8) are also important. These terms lead to a change in the form of the wave packet and hence to a distortion of the signal.

With so-called anomalous dispersion v_g may be greater than c if one calculates the group velocity from (12.10). One apparently encounters a contradiction to the theory of special relativity, according to which the signal velocity cannot be greater than the velocity of light in the vacuum. In the case of anomalous dispersion, however, the above mathematical description of the propagation of light no longer applies. Anomalous dispersion arises from the coupling of the light propagation and resonant oscillations of matter. The light is absorbed by the resonator and reemitted with a time delay. Our above description of the propagation of light is too simple to reproduce this correctly.

In quantum mechanics the motion in the positive x-direction of a free particle is described by the complex wave function

$$\psi(x,t) = e^{i(kx-\omega t)} \quad . \tag{12.12}$$

As a function of the wave number, the angular frequency is

$$\omega(k) = \frac{\hbar}{2m}k^2 \quad , \tag{12.13}$$

where m is the mass of the particle and $\hbar$ the Planck constant divided by 2π.

If we form a wave packet by superposition of plane waves of the form (12.12) in a narrow range about the mean wave number k_0, then we obtain from (12.10) the group velocity

$$v_g = \frac{\hbar}{m}k_0 \quad . \tag{12.14}$$

It is equal to the quantum mechanical expectation value of the velocity of the particle.

In the present chapter we shall not be concerned any further with quantum mechanics and we shall also steer clear of the propagation of light through matter. We require an example from classical physics and choose the propagation of water waves.

The theory of water waves is treated by A. Sommerfeld in *Vorlesungen über Theoretische Physik* (Lectures on Theoretical Physics) [12.1]. Various cases are considered: plane waves in deep water, in fairly deep water and in shallow water, where there are also differences between gravity waves and capillary waves. The latter are very short water waves in which the forces of surface tension are dominant. As a result of the non-trivial mathematical analysis we obtain the relation

$$\omega(k) = \sqrt{g|k|} \quad , \tag{12.15}$$

where g is the acceleration due to gravity; the relation holds for gravity waves in deep water. For the phase velocity it follows that

$$v_{\rm p} = \frac{\omega}{k} = \sqrt{\frac{g}{|k|}} = \sqrt{\frac{g\lambda}{2\pi}} \quad , \tag{12.16}$$

and for the group velocity

$$v_{\rm g} = \frac{d\omega}{dk} = \frac{1}{2}\sqrt{\frac{g}{|k|}} = \frac{1}{2}v_{\rm p} \quad . \tag{12.17}$$

The group velocity is equal to half the phase velocity. Phase and group velocities increase with the square root of the (mean) wavelength. Long waves on the sea move with the velocity of large ships, whereas short waves on a lake scarcely keep pace with a sailing-boat. That the group velocity is less than the phase velocity can be seen when one throws a stone into the water and observes the spreading ripples. If one watches sufficiently closely one sees that the wave crests start at the inner edge of the ring, travel through it and disappear again at the outer edge. The annular wave packet also changes its form. From an initial 3 or 4 wave crests it gradually becomes 5, 6 or more wave crests.

We shall follow this process in more detail with the computer and limit ourselves again to wave propagation in one direction. We limit ourselves to solutions with positive parity, i.e. to wave packets which are at all times symmetrical with respect to the origin. At time $t = 0$ we specify a wave packet modulated in Gaussian form,

$$\psi(x,0) = \exp\left(-\frac{x^2}{a^2}\right) \cos(k_0 x) \quad , \tag{12.18}$$

where k_0 is the mean wave number and

$$\lambda_0 = 2\pi/k_0 \quad . \tag{12.19}$$

the mean wavelength. For $a = 6\,{\rm m}$, $\lambda_0 = 10\,{\rm m}$ such a wave packet is shown in Fig. 12.1. We shall write $\psi(x,0)$ in the form of (12.4) and then proceed to times $t > 0$. Using the Fourier transformation we obtain the superposition amplitude $a(k)$,

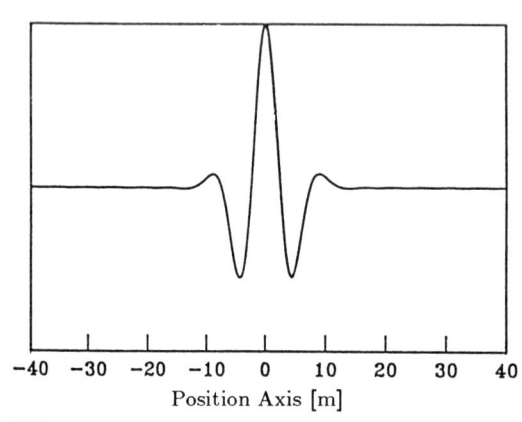

Fig. 12.1. Example of a wave packet of the form (12.18)

$$a(k) = \frac{2}{\pi} \int_0^\infty \exp\left(-\frac{x^2}{a^2}\right) \cos(k_0 x) \cos(kx) dx$$

$$= \frac{a}{2\sqrt{\pi}} \left\{ \exp\left(-\frac{(k+k_0)^2 a^2}{4}\right) + \exp\left(-\frac{(k-k_0)^2 a^2}{4}\right) \right\} . \quad (12.20)$$

It consists of the sum of two Gauss functions. Correspondingly with the substitutions in (12.4) we obtain the sum of two wave packets:

$$\psi(x,t) = \psi_1(x,t) + \psi_2(x,t) , \quad (12.21)$$

$$\psi_1 = \frac{a}{2\sqrt{\pi}} \int_{-\infty}^\infty \exp\left(-\frac{(k+k_0)^2 a^2}{4}\right) \cos(kx - \sqrt{g|k|}t) dk , \quad (12.22)$$

$$\psi_2 = \frac{a}{2\sqrt{\pi}} \int_{-\infty}^\infty \exp\left(-\frac{(k-k_0)^2 a^2}{4}\right) \cos(kx - \sqrt{g|k|}t) dk . \quad (12.23)$$

For a sufficiently large value of k_0 the first wave packet moves in the direction of the negative x-axis, the second in the direction of the positive x-axis. This corresponds to the ripples which we obtain when we throw a stone into the water: an initial disturbance of the water surface propagates in all directions. In this chapter we shall follow the motion of one of the two wave packets and shall notice how the wave crests overtake the wave packet and how the wave packet changes form in the course of time. For this it is sufficient to display $\psi_2(x,t)$ on the screen as a function of x for various values of t. Interested readers are recommended to calculate also the sum of the two wave packets and display on the screen how the wave packets split.

12.2 Numerical Method

In what follows we shall be concerned only with the calculation of $\psi_2(x,t)$ and therefore shall omit the index 2. We calculate the Fourier integral (12.23) by numerical integration, see Chap. 3.

The integral extends from $-\infty$ to $+\infty$. The integrand decreases rapidly, however, as soon as $(k-k_0)^2 a^2$ is large compared to 1. It suffices to limit the integral to the region $-4/a \leq k - k_0 \leq 4/a$. We then obtain

$$\psi(x,t) \approx \frac{a}{2\sqrt{\pi}} \int_{k_0-4/a}^{k_0+4/a} \exp\left(-\frac{(k-k_0)^2 a^2}{4}\right) \cos(kx - \sqrt{g|k|}t) dk. \quad (12.24)$$

The integrand contains several transcendental functions and has to be computed for every x-value and every time point. It would therefore be desirable to use an integration formula for (12.24) which required only a few mesh points, such as, for example, the Gauss-Legendre integration. This integration formula, however, is not very suitable for integrands which oscillate frequently. We therefore fall back on the well known trapezoidal rule (3.7) with equidistant meshes. Since the integrand almost vanishes at the ends of the interval, the Simpson rule would offer no advantage, see Chap. 3. We can apply the trapezoidal rule and use the integration weight Δk in the interior of the interval and also at its ends. We thus obtain the approximation

$$\psi(x,t) \approx \sum_{j=-m}^{m} a_j \cos(k_j x - b_j) \Delta k \quad , \tag{12.25}$$

where we employ the following abbreviations:

$$k_j = k_0 + j\Delta k \quad , \quad \Delta k = \frac{4}{am} \quad , \tag{12.26}$$

$$a_j = \frac{a}{2\sqrt{\pi}} \exp\left(-\frac{(k_j - k_0)^2 a^2}{4}\right) \quad , \quad b_j = \sqrt{g|k_j|}\, t \quad . \tag{12.27}$$

The number of integration points is $2m+1$. The indexing chosen here suggests itself because k_0 lies in the middle of the integration interval.

We shall use (12.25) in order to calculate the wave function $\psi(x,t)$ at a specified time t at the mesh points in the interval $[0, x_{\max}]$

$$x_i = ih \quad , \quad i = 0, \ldots, i_{\max} \quad , \quad i_{\max} = x_{\max}/h \quad . \tag{12.28}$$

In order to save computation time, we shall not calculate the cosine function in (12.25) at each mesh point, but use a recursion formula. We accordingly compare (12.25) at two neighbouring mesh points:

$$\psi(x_i,t) \approx \sum_{j=-m}^{m} a_j \cos(k_j x_i - b_j) \Delta k \quad , \tag{12.29a}$$

$$\psi(x_{i+1},t) \approx \sum_{j=-m}^{m} a_j \cos(k_j x_{i+1} - b_j) \Delta k$$

$$= \sum_{j=-m}^{m} a_j \cos(k_j(x_i + h) - b_j) \Delta k \quad . \tag{12.29b}$$

Now quite generally we have:

$$\cos(x+h) = \cos x \cos h - \sin x \sin h \quad ,$$
$$\sin(x+h) = \sin x \cos h + \cos x \sin h \quad , \tag{12.30}$$

from which it follows that

$$\cos(k_j x_{i+1} - b_j) = \cos(k_j x_i - b_j) \cos(k_j h) - \sin(k_j x_i - b_j) \sin(k_j h) \quad ,$$
$$\sin(k_j x_{i+1} - b_j) = \sin(k_j x_i - b_j) \cos(k_j h) + \cos(k_j x_i - b_j) \sin(k_j h) \quad . \tag{12.31}$$

Accordingly we can start out from $\cos(k_j x_0 - b_j)$, $\sin(k_j x_0 - b_j)$, $\cos(k_j h)$

and $\sin(k_j h)$, and hence determine all subsequent $\cos(k_j x_i - b_j)$ from the recursion formula (12.31).

Many recursion formulae are not useful numerically because of the accumulation of rounding errors. A small test shows, however, that the recursion formulae (12.30) are satisfactory in this respect. If one specifies $\cos(0.01)$ and $\sin(0.01)$ and applies (12.30) 10 000 times in order to calculate $\cos(100)$, the rounding error with DOUBLE PRECISION is less than 10^{-10}.

12.3 Programming

The programming of (12.25) together with the recursion formulae (12.31) is so simple that we carry it all out in the main program KAP12 (see Fig. 12.3; the notation employed is presented in Fig. 12.2).

Physical title	FORTRAN title	Physical title	FORTRAN title
π	PI	i_{max}	IMAX
g	G	h	H
a	A	x_i	X(I)
b	B	$\psi(x_i, t)$	PSI(I)
λ_0	LAMBDA	k_j	KJ
k_0	KO	a_j	AJ
Δk	DELTAK	b_j	BJ
τ	TAU	m	M
v_p	VP	$a_j \Delta k \cos(k_j x_i - b_j)$	COSKX
v_g	VG	$a_j \Delta k \sin(k_j x_i - b_j)$	SINKX
t	T	$\cos(k_j h)$	COSKH
		$\sin(k_j h)$	SINKH

Fig. 12.2. Notation employed in main program KAP12

In KAP12 the quantities m, i_{max} and h are fixed constants (line 103). The mean wavelength λ_0 is input, together with the half-width b of the wave packet, which is related to a by the equation

$$b = 2a\sqrt{\ln 2} \tag{12.32}$$

(see line 106). In lines 107 to 111 k_0 is determined from (12.19), the period τ from (12.2) and (12.15), and v_p and v_g from (12.16) and (12.17). The x_i are calculated from (12.28) in the DO-loop starting at line 113. In order to evaluate (12.25) for all x_i, the $\psi(x_i,t)$ are first set to zero (line 117). In the nest of DO-loops from line 119 to line 133 there follows the summation over all j. First k_j, a_j and b_j are determined from (12.26) and (12.27) (lines 120–122). Subse-

```
          PROGRAM KAP12                                              100

          IMPLICIT DOUBLE PRECISION (A-H,O-Z)                        101
          DOUBLE PRECISION KO, KJ, LAMBDA                            102
          PARAMETER (M=100, IMAX=250, H=0.16D0)                      103
          PARAMETER (G=9.81D0, PI=3.1415926536D0)                    104
          DIMENSION PSI(0:IMAX), X(0:IMAX)                           105

    *     (Input: LAMBDA, B)

          A=B/(2.D0*SQRT(LOG(2.D0)))                                 106
          K0=2.D0*PI/LAMBDA                                          107
          DELTAK=4.D0/(A*M)                                          108
          TAU=SQRT(2.D0*PI*LAMBDA/G)                                 109
          VP=SQRT(G*LAMBDA/(2.D0*PI))                                110
          VG=VP/2.D0                                                 111
          T=0.D0                                                     112
          DO 10 I=0,IMAX                                             113
            X(I)=I*H                                                 114
    10    CONTINUE                                                   115

    *     (Output: TAU, VP, VG)

          DO 20 I=0,IMAX                                             116
            PSI(I)=0.D0                                              117
    20    CONTINUE                                                   118
          DO 40 J=-M,M                                               119
            KJ=K0+J*DELTAK                                           120
            AJ=A/(2.D0*SQRT(PI))*EXP(-((KJ-K0)*A)**2/4.D0)           121
            BJ=SQRT(G*ABS(KJ))*T                                     122
            COSKX=COS(KJ*X(0)-BJ)*AJ*DELTAK                          123
            SINKX=SIN(KJ*X(0)-BJ)*AJ*DELTAK                          124
            COSKH=COS(KJ*H)                                          125
            SINKH=SIN(KJ*H)                                          126
            DO 30 I=0,IMAX                                           127
              PSI(I)=PSI(I)+COSKX                                    128
              COSII=COSKX*COSKH-SINKX*SINKH                          129
              SINKX=SINKX*COSKH+COSKX*SINKH                          130
              COSKX=COSII                                            131
    30      CONTINUE                                                 132
    40    CONTINUE                                                   133

    *     (Output: PSI)
    *     (Input: new T)

          END                                                        134
```

Fig. 12.3. Numerical portion of the main program KAP12

quently the quantities $\cos(k_j x_0 - b_j)$, $\sin(k_j x_0 - b_j)$, $\cos(k_j h)$ and $\sin(k_j h)$ are computed, which are needed as starting values for the recursion (12.31) (lines 123–126). Here we multiply the values $\cos(k_j x_0 - b_j)$ and $\sin(k_j x_0 - b_j)$ by the factor $a_j \Delta k$, which occurs in the evaluation of (12.25). In the final DO-loop a summation step from (12.25) is first carried out (line 128), then the quantities $a_j \Delta k \cos(k_j x_{i+1} - b_j)$ and $a_j \Delta k \sin(k_j x_{i+1} - b_j)$ are determined (lines 129–131).

First of all the wave functions for $t = 0$ are computed and output graphically. Then a new value for the time t can be input. As we must not leave the graphics mode for this input, we use the FUNCTION subprogram GVALUE. GVALUE supplies a function value of DOUBLE PRECISION type and is called by

T = GVALUE(NOTE)   .

Here NOTE denotes the CHARACTER expression for a text which is output in the top margin of the screen. After calling GVALUE one can input a number, which is passed back as the function value to the calling program.

In order better to be able to compare the wavefunctions for different times with one another, we show three wavefunctions, one below the other, each in a different graphics window. The subroutine GWINDO enables us to define the smallest and the largest value on each axis, together with the extent of each graphics window. GWINDO is called in the following way:

CALL GWINDO(XMIN, XMAX, YMIN, YMAX, XWMIN, XWMAX, YWMIN, YWMAX)

Here XMIN and XMAX are the smallest and the largest values on the x-axis, YMIN and YMAX the smallest and largest values on the y-axis. They thus correspond to the parameters XMIN, XMAX, YMIN, and YMAX of the subroutine GOPEN (see Chap. 4). The parameters XWMIN and XWMAX denote the left and right-hand boundaries of the graphics window in the screen. They are DOUBLE PRECISION quantities between 0 and 100. The value XWMIN=0 indicates that the left-hand edge of the graphics window coincides with the left-hand edge of the screen, the value XWMIN=50 that the left-hand edge of the graphics window lies in the centre of the screen. In the same way YWMAX and YWMIN indicate the top and bottom edges of the graphics window. After GWINDO has been called the quantities X and Y in GMOVE, GDRAW, GMARK, GBOX, etc. refer to the newly defined axes, and the output relates to the portion of the screen delimited by XWMIN, XWMAX, YWMIN and YWMAX. Using GWINDO we can accordingly display several diagrams in separate regions of the screen. GWINDO serves only to define the new graphics window, not to draw the frame, the axes or the grid. For that one uses the subroutine GCHART, which is called by

CALL GCHART(XTICK, XCROSS, YTICK, YCROSS,

   XNAME, YNAME, IGRID, ICROSS, IWHERE)   .

Most of the parameters of GCHART are already familiar to us from GOPEN. XTICK and YTICK indicate the required distances along the x- and y-axes for marks. The CHARACTER expressions XNAME and YNAME contain the text with which the axes are to be labelled. If only a blank is given for XNAME or YNAME, the corresponding axis will have no label and also the marks on this axis will not be labelled. IGRID and ICROSS indicate whether a grid and axes are to be shown. The intersection of the axes can be specified by XCROSS and YCROSS: in GOPEN this was always the point (0,0). Using the parameter IWHERE one can specify on which side the graphics shall be labelled. IWHERE=1 causes the label to appear below and to the left, and hence in the same place as in GOPEN. With IWHERE=2 one places the label below and to the right, with IWHERE=3 above and to the right, and with IWHERE=4 above and to the left.

The complete program here described in summary is to be found on the diskette in the file KAP-12.FOR.

12.4 Exercises

12.4.1 Using the wavefunctions displayed on the screen, determine the period, the phase and group velocities. Compare the values so determined with those calculated from (12.15) to (12.17). Are there values for the wavelength and the half-width of the wave packet, for which the group velocity is not a physically meaningful concept?

12.4.2 Study the variation with time of the form of a wave packet. Which wave packets are rather stable, and which change their shape particularly quickly? Can one conclude from the time evolution of a single wave packet that long water waves travel faster?

12.5 Solutions to the Exercises

12.5.1 The period can be determined most simply by finding the times at which the maxima and the nodes of the wave function lie in the same positions as at time $t = 0$. By dividing the wavelength by the period one obtains the phase velocity. The group velocity is determined by establishing the speed of the place in the wave packet where the maxima are highest. The determination of a group velocity is meaningful only if the neglect of the quadratic term in (12.8) is at least approximately justifiable. This is the case when the wavelength is significantly shorter than the half-width of the wave packet.

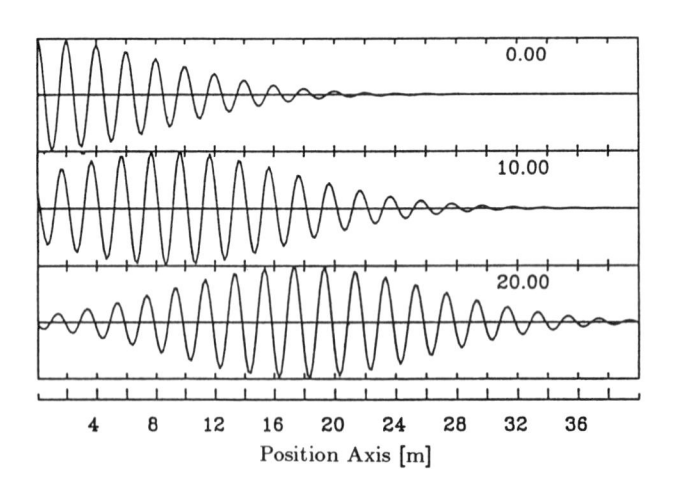

Fig. 12.4. Wave packet with $\lambda_0 = 2\,\mathrm{m}$, $b = 20\,\mathrm{m}$ after 0, 10 and 20 s

12.5.2 Wave packets in which the average wavelength is much shorter than the half-width, i.e. which form a long wave train, change their form only slowly (see Fig. 12.4). The reason for this is that, according to (12.20), for such wave packets $a(k)$ is significantly different from zero only in a small region about k_0: the wave packet is approximately "monochromatic".

On the other hand wave packets change their form rapidly if the half-width is smaller than or equal to the average wavelength, i.e. if it involves only a very short wave train, see

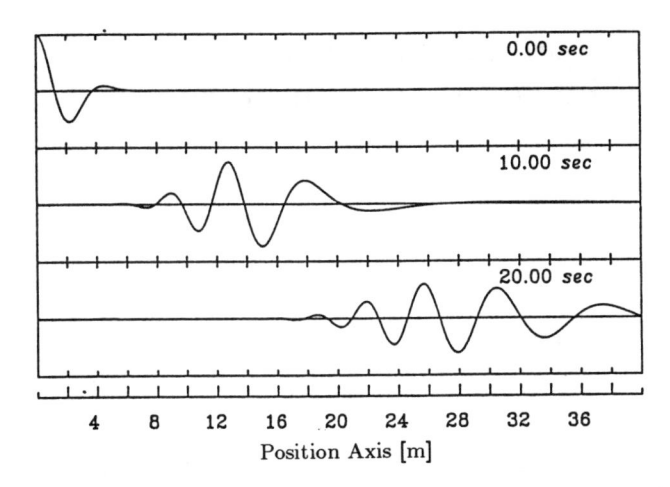

Fig. 12.5. Wave packet with $\lambda_0 = 5\,\text{m}$, $b = 5\,\text{m}$ after 0, 10 and 20 s

Fig. 12.5. In short wave packets one can clearly recognise that after some time the wavelength at the front end of the wave packet is longer than at the rear end (see Fig. 12.5 again). This results from the fact that the long wave components of such a wave packet move faster than the short wave components and are consequently to be found at the front end of the packet.

13. Solution of the Radial Schrödinger Equation by the Fox-Goodwin Method

13.1 Formulation of the Problem

Solving the Schrödinger equation is the central problem of non-relativistic quantum mechanics. A simple case is the study of the motion of a particle without spin in an external potential. The time-independent Schrödinger equation in this case reads

$$H\psi(\mathbf{r}) = E\psi(\mathbf{r}) \; , \tag{13.1a}$$

with the Hamilton operator

$$H = -\frac{\hbar^2}{2m}\Delta + V(\mathbf{r}) \; . \tag{13.1b}$$

If the particle is scattered by the potential, the energy E may be given any positive value. If the particle is bound by the potential, E becomes negative and can only take discrete values.

Some complicated problems can be reduced to solving the one-particle Schrödinger equation. Solutions of the one-particle Schrödinger equation accordingly form the basis of states of the shell models in atomic physics and nuclear physics. The mutual scattering of two particles also leads to a Schrödinger equation of type (13.1). In this case $\mathbf{r}$ is the separation vector of the two particles, m is the reduced mass, and E is the energy in the centre of mass system, i.e. in that inertial system in which the centre of mass of the two particles is at rest.

Analytic solutions of (13.1) are known only for a few special forms of potential. In general the equation has to be solved numerically. It is fairly easy to do this for a spherically symmetric potential, i.e. when

$$V(\mathbf{r}) = V(r) \; . \tag{13.2}$$

We shall assume in what follows that this condition is fulfilled.

Restriction to central potentials makes the introduction of spherical coordinates (r, θ, φ) appropriate. The Schrödinger equation is then separable, i.e. one obtains one equation for the radial motion and one for the angular motion. We shall consider the process of separation in more detail in Chap. 17.

The differential equation for the angular motion does not involve the potential $V(r)$. Its solutions are the spherical harmonics $Y_{lm}(\theta, \varphi)$. The differential equation for the radial motion reads

$$-\frac{\hbar^2}{2m}\frac{d^2}{dr^2}u_l(r) + \frac{\hbar^2}{2m}\frac{l(l+1)}{r^2}u_l(r) + V(r)u_l(r) = Eu_l(r) \quad . \tag{13.3}$$

Using the spherical harmonics $Y_{l,m}(\theta,\varphi)$ and the radial functions $u_l(r)$, the general solution $\psi(\boldsymbol{r})$ of the Schrödinger equation (13.1) can be expressed as a superposition of partial waves:

$$\psi(\boldsymbol{r}) = \sum_{l,m} c_{l,m}\frac{1}{r}u_l(r)Y_{l,m}(\vartheta,\varphi) \quad . \tag{13.4}$$

The Schrödinger equation (13.1) is satisfied by each partial wave, and consequently by their sum.

The quantity l occurring in (13.3) and (13.4) is the angular momentum quantum number. The symbol m is generally used in quantum mechanics both for the mass and also for the magnetic quantum number (quantum number of the z-component of the angular momentum). In (13.4) m is the magnetic quantum number.

The decomposition of the total wave function into angular momentum partial waves brings certain advantages. Bound states contain only one partial wave. For scattering states one must take account of more than one partial wave, but usually only a small number. The reason for this is as follows. Apart from the Coulomb potential most physical potentials are of short range. Particles with finite energy and large angular momentum cannot enter into the interaction region. In the radial equation (13.3) this is expressed by the second term, which contains the centrifugal barrier $\hbar^2 l(l+1)/(2mr^2)$. The centrifugal barrier acts like a potential which is so strong for large values of l that $V(r)$ becomes negligible by comparison. When $V(r)$ is neglected equation (13.3) becomes the free Schrödinger equation, whose analytic solutions are known. They are treated in Chap. 18. We therefore have to solve the radial equation (13.3) for only a few values of l.

The radial Schrödinger equation (13.3) is a differential equation of the second order. Accordingly one needs two boundary conditions to get a unique solution.

One of the conditions is obtained from the requirement that the probability density $|\psi(\boldsymbol{r})|^2$ must remain finite at the origin of coordinates. The probability density is given by

$$|\psi(\boldsymbol{r})|^2 = \left|\sum_{l,m}\frac{u_l(r)}{r}c_{l,m}Y_{l,m}(\vartheta,\varphi)\right|^2 \quad . \tag{13.5}$$

The probability density can remain finite at the origin for arbitrary expansion coefficients $c_{l,m}$ only if

$$u_l(0) = 0 \tag{13.6}$$

This is our first boundary condition.

The second boundary condition arises in connection with the normalisation of the wave functions $u_l(r)$. We must distinguish here between bound state solutions ($E<0$) and scattering solutions ($E>0$).

In bound states the motion of the particle (or the relative motion of the two particles) is restricted to a finite space, i.e. the wave function must be normalisable to unity. As the second boundary condition we can tentatively set the wave function equal to any chosen number, e.g. unity, at an arbitrarily chosen distance $r_0(\neq 0)$:

$$u_l(r_0) = 1 \ . \tag{13.7}$$

The solution of the radial equation is thereby determined. We shall find it numerically and see that, for values of r greater than the range of the potential, it has the form

$$u_l(r) \to a_l e^{-\kappa r} + b_l e^{\kappa r} \tag{13.8}$$

with

$$\kappa = \left(\frac{2m(-E)}{\hbar^2}\right)^{1/2} \ . \tag{13.9}$$

The condition for normalisability,

$$\int_0^\infty |u_l(r)|^2 dr = \text{finite} \ , \tag{13.10}$$

is, however, fulfilled only if b_l vanishes, since otherwise the function $u_l(r)$ would increase without limit for large values of r. We shall see that b_l is a function of the energy E introduced by us into the radial equation. For a binding energy E_i, b_l will pass through zero as a function of E:

$$b_l(E_i) = 0 \ . \tag{13.11}$$

We shall recognise this zero from the fact that for large values of r the wave function $u_l(E, r)$ flips between large negative values and large positive values for slight variations in the energy E. The energy at which the change occurs is one of the binding energies E_i. For $E = E_i$ the wave function $u_l(E, r)$ becomes normalisable. For practical computation we must replace the upper integration limit in the integral (13.10) by a value R which is not excessively large, since our solution becomes numerically unstable for large r-values. Because of the arbitrary second boundary condition (13.7) the bound state function $u_l(E_i, r)$ obtained is normalised, but not to unity. We can, however, compute the integral

$$\int_0^R |u_l(E_i, r)|^2 dr = N^2 \tag{13.12}$$

and then with

$$\hat{u}_l(E_i, r) = \frac{1}{N} u_l(E_i, r) \tag{13.13}$$

we get a bound state function normalised to unity. Because of the linearity of the Schrödinger equation, also $\hat{u}_l(E_i, r)$ satisfies the equation.

For scattering solutions the energy E has the physical significance of an energy imparted by an accelerator, and may take any positive value. With the boundary conditions (13.6) and (13.7) we again obtain a uniquely defined solution $u_l(E, r)$ which now, however, does not vanish asymptotically. If the potential $V(r)$ has no long range component, then $u_l(E, r)$ has the asymptotic form

$$u_l(E, r) \to N \sin\left(kr - \frac{l\pi}{2} + \delta_l\right) \tag{13.14}$$

with

$$k = \left(\frac{2mE}{\hbar^2}\right)^{1/2}. \tag{13.15}$$

The second boundary condition (13.7) will give us some value for N. If we wish, we can normalise $u_l(E, r)$ by multiplying it by $1/N$. Normalisation means in this case that the wave has an asymptotic amplitude equal to unity. The quantity δ_l is called the scattering phase. It is a function of the energy E and will be discussed in more detail in Chaps. 18 and 19. The quantity $l\pi/2$ is a phase shift produced by the angular momentum barrier.

As a physical example we shall consider the mutual scattering of two α-particles with the Coulomb interaction switched off. The α-α scattering is rather interesting, since α-particles are in reality not point masses, but the nuclei of helium atoms. They consist of two protons and two neutrons each. If two α-particles are so close to one another that they interpenetrate, then the Pauli principle has a strong influence on the interaction. Nevertheless the mutual scattering of two α-particles can be described very well by means of a simple potential. For the potential we use the expression [13.1]

$$V(r) = -V_0 \exp(-\beta r^2) \tag{13.16a}$$

with

$$V_0 = 122.694\,\text{MeV} \quad \text{and} \quad \beta = 0.22\,\text{fm}^{-2}. \tag{13.16b}$$

For a realistic computation one would also have to take account of the Coulomb potential.

We shall solve the radial Schrödinger equation (13.3) with the potential (13.16) and examine the solutions in detail. As angular momentum quantum numbers we choose $l = 0, 2, 4$ and 6. Odd angular momentum quantum numbers are forbidden for two α-particles by the Pauli principle. Our unit of energy is 1 MeV, the unit of length is 1 fm.

In the region of negative energies we shall find four bound states, namely three bound states for $l = 0$ and one for $l = 2$. The two lowest bound states for $l = 0$ and the bound state for $l = 2$ are unphysical. They are Pauli-forbidden, as they correspond to too dense packing of the 8 nucleons. Just because of these unphysical bound states the potential (13.16) reproduces the scattering wave functions well, for these are orthogonal to the bound state functions. The third bound state for $l = 0$ is also unphysical, but for a different reason. We have omitted the Coulomb interaction, and thereby arises a weakly bound state of two α-particles. When the Coulomb potential is included this bound state becomes the well known sharp resonance at 0.092 MeV, which one can also regard as the short-lived ground state of the ^{8}Be nucleus.

Since we are dealing with a relative motion, we must use the reduced mass of two α-particles instead of the mass of one α-particle in the radial equation (13.3), and also in (13.9) and (13.15). It occurs only in the combination $\hbar^2/2m$. We use the value

$$\frac{\hbar^2}{2m} = 10.375 \, \text{MeV fm}^2 \quad . \tag{13.17}$$

13.2 Numerical Method of Solution

The radial Schrödinger equation (13.3) is equivalent to

$$u''(r) + w(r)u(r) = 0 \quad , \tag{13.18}$$

where we have used the abbreviation

$$w(r) = \frac{2m}{\hbar^2}(E - V(r)) - \frac{l(l+1)}{r^2} \quad . \tag{13.19}$$

The index l will from now on be omitted.

For the numerical treatment of (13.18), the derivative $u''(r)$ will be replaced by a finite difference expression. We obtain the simplest approximation by using the three point formula (2.8), neglecting the error term $O(h^2)$

$$u''(r) = \frac{u(r+h) + u(r-h) - 2u(r)}{h^2} \quad . \tag{13.20}$$

Equation (13.18) then becomes

$$\frac{u(r+h) + u(r-h) - 2u(r)}{h^2} + w(r)u(r) = 0 \quad , \tag{13.21}$$

or

$$u(r+h) = 2u(r) - u(r-h) - h^2 w(r)u(r) \quad . \tag{13.22}$$

If one now postulates, for example,

$$u(0) = 0 \tag{13.23}$$

and

$$u(h) = 1 \tag{13.24}$$

then using (13.22) one can first of all determine $u(2h)$. From $u(h)$ and $u(2h)$ follows $u(3h)$, and so on. One thus obtains $u(r)$ at a grid of equidistant points, with mesh width h.

The approximate equation (13.22) can be improved, by taking the error term occurring in (2.8) into account in a convenient way. From (2.8) and (2.9) we obtain

$$\frac{u(r+h)+u(r-h)-2u(r)}{h^2} = u''(r) + \frac{h^2}{12}u^{(4)}(r) + O(h^4) \quad . \tag{13.25}$$

If we use (13.25) in order to replace $u''(r)$ in (13.18), the correction terms appear on the right-hand side:

$$\frac{u(r+h)+u(r-h)-2u(r)}{h^2} + w(r)u(r) = \frac{h^2}{12}u^{(4)}(r) + O(h^4) \quad . \tag{13.26}$$

There is an elegant method of causing the correction term of order h^2 to vanish, without having to abandon the three-point formula. This method is called the Fox-Goodwin, Noumerov, or Cowell method [13.2]. We shall call it the Fox-Goodwin method.

The trick of the method lies in applying to (13.18) the operator

$$1 + \frac{h^2}{12}\frac{d^2}{dr^2} \tag{13.27}$$

thus obtaining

$$\left(1 + \frac{h^2}{12}\frac{d^2}{dr^2}\right)u''(r) + w(r)u(r) + \frac{h^2}{12}\frac{d^2}{dr^2}(w(r)u(r)) = 0 \quad . \tag{13.28}$$

One then uses equation (13.25) to substitute for the first term, and gets

$$\frac{u(r+h)+u(r-h)-2u(r)}{h^2} + O(h^4) + w(r)u(r)$$
$$+ \frac{h^2}{12}\frac{d^2}{dr^2}(w(r)u(r)) = 0 \quad . \tag{13.29}$$

All terms of order h^4 and higher are then neglected. One can now replace the second derivative of $(u(r)w(r))$ by an approximation of type (13.20), because the factor $h^2/12$ already occurring in (13.29) now makes the first correction term of order h^4. We therefore obtain

$$\frac{u(r+h)+u(r-h)-2u(r)}{h^2} + w(r)u(r)$$
$$+ \frac{w(r+h)u(r+h)+w(r-h)u(r-h)-2w(r)u(r)}{12} = 0 \quad , \tag{13.30}$$

or

$$u(r+h) = \frac{2u(r) - u(r-h) - (h^2/12)(10w(r)u(r) + w(r-h)u(r-h))}{1 + (h^2/12)w(r+h)} \quad . \tag{13.31}$$

This equation now takes the place of (13.22) and again allows us to compute the function $u(r)$ on an equidistant grid of points, with mesh width h. The neglected terms in (13.31) are two orders of h smaller than those neglected in (13.22).

13.3 Programming

In order to solve (13.3) using the Fox-Goodwin recursion (13.31) we write a subroutine entitled FOX, which solves the differential equation (13.18).

We introduce the arrays (r_i), (u_i) and (w_i), with

$$r_i = ih \quad , \quad u_i = u(r_i) \quad , \quad w_i = w(r_i) \quad , \quad i = 0, 1, \ldots, i_b \quad . \quad (13.32)$$

The radial wave function is to be computed in the interval $[0, b]$, i.e. the lower interval boundary lies at $r = 0$, the upper at $r = b$. The interval $[0, b]$ is divided into i_b subintervals of length

$$h = \frac{b}{i_b} \quad . \quad (13.33)$$

The number of mesh points is therefore equal to $(i_b + 1)$.

With the notation of (13.32) and (13.33), (13.31) can be rewritten in the following form:

$$u_{i+1} = \frac{2u_i - u_{i-1} - (h^2/12)(10 w_i u_i + w_{i-1} u_{i-1})}{1 + (h^2/12) w_{i+1}} \quad . \quad (13.34)$$

Physical title	FORTRAN title	Physical title	FORTRAN title
i_b	IB	u_i	U(I)
b	B	w_i	W(I)
h	H		

Fig. 13.1. Notation used in the subroutine FOX

Figure 13.2 shows the subroutine FOX; the notation used is to be found in Fig. 13.1. The calling program transfers to the subroutine FOX the upper boundary b of the integration interval, the number of integration steps i_b and the values w_i of the function $w(r)$. It receives back from FOX the values u_i of the solution function $u(r)$ (see parameter list in line 100). In lines 104 and 105 the

```
      SUBROUTINE FOX(IB,B,W,U)                                        100

      IMPLICIT DOUBLE PRECISION (A-H,O-Z)                             101
      DIMENSION W(0:IB), U(0:IB)                                      102
      H=B/IB                                                          103
      U(0)=0.D0                                                       104
      U(1)=1.D0                                                       105
      DO 10 I=1,IB-1                                                  106
         U(I+1)=(2.D0*U(I)-U(I-1)-H*H/12.D0*(10.D0*W(I)*U(I)+W(I-1)   107
     &          *U(I-1)))/(1.D0+H*H/12.D0*W(I+1))                     108
10    CONTINUE                                                        109
      END                                                             110
```

Fig. 13.2. Subroutine FOX

Physical title	FORTRAN title	Physical title	FORTRAN title
V_0	V0	r	R
β	BETA		

Fig. 13.3. Notation used in the FUNCTION subprogram V

```
      DOUBLE PRECISION FUNCTION V(R)                        200
      IMPLICIT DOUBLE PRECISION (A-H,O-Z)                    201
      PARAMETER (V0=122.694D0, BETA=0.22D0)                  202
      PARAMETER (EXPMAX=50.D0)                               203
      IF(BETA*R*R.LE.EXPMAX) THEN                            204
         V=-V0*EXP(-BETA*R*R)                                205
      ELSE                                                   206
         V=0.D0                                              207
      ENDIF                                                  208
      END                                                    209
```

Fig. 13.4. FUNCTION subprogram V

boundary conditions (13.23) and (13.24) are specified. The recursion (13.34) is carried out in the DO-loop from line 106.

In order to be able to change the potential function easily, we introduce the FUNCTION subprogram V, see Fig. 13.3 and Fig. 13.4. In our case V calculates the α-α interaction potential for the separation distance r from (13.16) (line 205). The potential parameters are fixed (line 202). The IF-block of lines 204 to 208 prevents a possible exponent overflow and also ensures that the exponential function is not computed unnecessarily often.

Figure 13.6 shows the numerical portion of the main program KAP13; the notation used is listed in Fig. 13.5. In the DO-loop from line 307 the mesh points r_i are computed from (13.32), and the function values $w(r_i) = w_i$ from (13.19). In line 306 we set $w_0 = 0$. This needs a justification since, for non-vanishing angular momentum, $w(r)$ diverges as $r \to 0$. In equation (13.18) $w(r)$ occurs only in the product $w(r)u(r)$, and we have set $u(0)$ equal to zero according to the boundary condition (13.23). The quantity "infinity times zero", however, is undefined. With our choice $w(0) = 0$ we arbitrarily set $w(0)u(0)$ equal to zero. The justification for this will be obtained numerically: whenever the angular momentum barrier dominates $V(r)$, our solution $u_l(r)$ transforms into the known analytic solution of the free Schrödinger equation (see Chap. 18). This also answers the question whether the neglect of the correction terms of order h^4 in equation (13.29) for $l \neq 0$ and small values of r was permissible. The numerical portion of KAP13 ends with the call of the subroutine FOX.

The value for $\hbar^2/(2m)$ is fixed by (13.17) (line 302), whereas the energy E, the angular momentum quantum number l, the upper boundary b of the desired integration interval $[0, b]$ and the number i_b of integration steps are input. As output one obtains the radial wave function $u_l(r)$ in graphical form and optionally also in numerical form.

The complete program is to be found on the diskette in file KAP-13.FOR.

Physical title	FORTRAN title	Physical title	FORTRAN title
$\dot{\imath}_b$	IB	$\hbar^2/2m$	H2M
b	B	l	L
h	H	E	E
r_i	R(I)	u_i	U(I)
		w_i	W(I)

Fig. 13.5. Notation used in main program KAP13

```
      PROGRAM KAP13                                              300

      IMPLICIT DOUBLE PRECISION (A-H,O-Z)                        301
      PARAMETER (IMAX=2000, H2M=10.375D0, BMAX=30.D0)            302
      DIMENSION R(0:IMAX), U(0:IMAX), W(0:IMAX)                  303
*     (Input: E,L,B,IB)

      H=B/IB                                                     304
      R(0)=0.D0                                                  305
      W(0)=0.D0                                                  306
      DO 10 I=1,IB                                               307
        R(I)=I*H                                                 308
        W(I)=(E-V(R(I)))/H2M-L*(L+1)/(R(I)*R(I))                 309
10    CONTINUE                                                   310
      CALL FOX   (IB,B,W,U)                                      311

*     (Output: U)

      END                                                        312
```

Fig. 13.6. Numerical portion of main program KAP13

13.4 Exercises

13.4.1 Test the accuracy of the Fox-Goodwin method in the following manner:

a) Investigate for $l = 0$ the stability of the zeros of the wave function for small number of integration steps. Carry out this investigation with energies in the region 1 MeV, 10 MeV and 30 MeV. *Hint:* choose the upper limits of integration so that apart from the origin they produce just three zeros on the screen. For the number of integration steps take the values 10, 20, 40, 80 and 160.

b) Investigate the stability of the zeros with a large number of integration steps. *Hint:* proceed as in a), and choose as the number of integration steps the values 500, 1000, 2000.

c) For $l \neq 0$ the error terms near the origin can become very large. Investigate whether the Fox-Goodwin method nevertheless leads to numerically stable solutions.

13.4.2 Find the bound states mentioned in Sect. 13.1 by determining the energies for which the radial wave function at large distances changes sign (the region between 7 fm and 10 fm is a good choice). Notice the short range of the three Pauli-forbidden states.

13.4.3 Compute scattering waves for energies between 0.01 MeV and 30 MeV:

a) For $l = 0$. Notice that the wave functions, more or less independently of the energy, have zeros at distances of about 0.9 fm and 2.0 fm. Can you give an explanation for this?

b) For $l = 2, 4, 6, \ldots$ Note particularly the behaviour of the wave functions in the neighbourhood of the origin.

13.4.4 a) Extend the subroutine V so that it computes the potential

$$V(r) = -V_0 \exp(-\beta r^2) + V_c(r) \tag{13.35}$$

where $V_c(r)$ describes the Coulomb interaction of two α-particles:

$$V_c(r) = \frac{4e^2}{r} \text{erf}(\gamma r) \quad . \tag{13.36}$$

Remark: The Coulomb potential $4e^2/r$ is valid only for point particles. The finite extent of the charge distribution in α-particles is approximately taken into account by the factor $\text{erf}(\gamma r)$, where the parameter γ is adjusted to the radius of the α-particle. Use the following numerical values [13.1]:

$$e = 1.2 \,(\text{MeV fm})^{1/2} \quad \text{and} \quad \gamma = 0.75 \,\text{fm}^{-1} \quad . \tag{13.37a,b}$$

A 6-figure approximation to the Gaussian error function $\text{erf}(x)$ is provided by [13.3]:

$$\text{erf}(x) = 1 - (a_1 t + a_2 t^2 + a_3 t^3 + a_4 t^4 + a_5 t^5) \exp(-x^2) \quad , \tag{13.38}$$

with

$$t = \frac{1}{1+px} \quad , \quad p = 0.3275911 \quad , \quad a_1 = 0.254829592 \quad , \quad a_2 = -0.284496736 \quad ,$$

$$a_3 = 1.421413741 \quad , \quad a_4 = -1.453152027 \quad , \quad a_5 = 1.061405429 \quad . \tag{13.39}$$

b) Investigate the wave function in the neighbourhood of the resonance energy at 0.092 MeV. The resonance is very sharp! Carry out the search for the resonance state in a similar manner to that used above for bound states. Raise `IMAX` in the `PARAMETER` instruction (line 302 of Fig. 13.6) to 10 000 and `BMAX` to 160.D0.

13.5 Solutions to the Exercises

13.5.1 a) For $l = 0$, $E = 10$ MeV, $b = 5$ fm and 10 integration intervals the third zero lies at 3.6711 fm. This value is found by linearly interpolating the numerical printout of the wave function between 3.5 and 4.0 fm. Raising the number of integration intervals to 20, 40, 80, 160 causes the position of the zero to converge very rapidly to 3.770122 fm. The error in the position of the third zero decreases, for each doubling of the number of integration intervals, by a factor of about 1/18 to 1/20.

What is surprising is the quality of the approximation with a very small number of integration intervals. Even with an interval length of 0.5 fm one still obtains a wave function which looks similar to the true one.

b) In Chap. 2 we have seen that the numerical computation of a second derivative by means of a second order difference can be risky. Because our personal computer in single precision works with only 4 bytes for a floating point number, we have so far put our programs in double precision, i.e. with 8 bytes per floating point number. This doubled precision, however, is now so high that even with 2000 integration steps for a half wavelength no loss of accuracy can be detected in the wave function.

c) With the relatively high angular momentum number $l = 10$ and $E = 20$ MeV the first zero lies at 10.828 fm. In spite of the apparently exotic behaviour of the wave function for small r-values (the function increases like r^{11}!) we obtain here also, even with a mesh width of 0.5 fm, wave functions which are similar to the true ones.

13.5.2 For $l = 0$ the lowest bound state lies at -76.9036145 MeV. Figure 13.7 shows it in the region 0 to 5 fm. The maximum lies at about 0.84 fm. The experimental radius of the α-particle amounts to about 1.6 fm, i.e. in this lowest bound state the two α-particles would lie so close together that they would overlap almost completely. It is clear that here the Pauli principle regarding the neutrons and protons must play a crucial role. As stated in Sect. 13.1, the state is Pauli-forbidden.

The second bound state with $l = 0$ has the energy $-29.00048\,\text{MeV}$ and is shown in Fig. 13.8. The wave function has a zero and is orthogonal to the wave function of the first bound state. The bound state with $l = 2$ is shown in Fig. 13.9. Notice the different type of behaviour of the wave function near the origin.

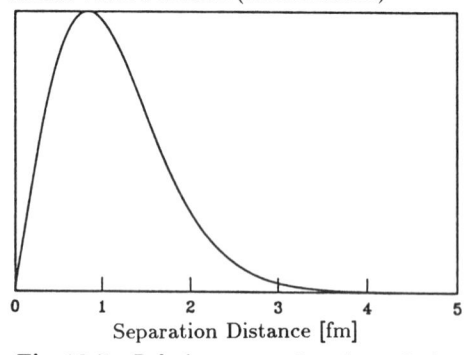

Fig. 13.7. Relative wave function of the lowest bound state of two α-particles (Pauli-forbidden)

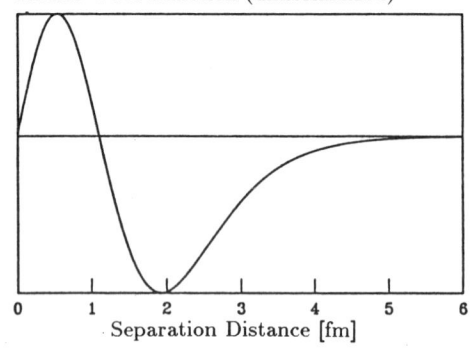

Fig. 13.8. Relative wave function of the second bound state of two α-particles (Pauli forbidden)

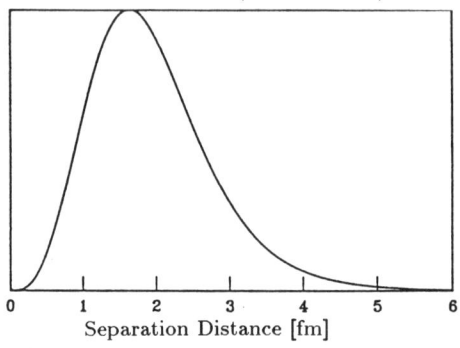

Fig. 13.9. Relative wave function of the Pauli-forbidden bound state of two α-particles with $l = 2$

13.5.3 a) A scattering wave for $l = 0$ and $1\,\text{MeV}$ is shown in Fig. 13.10; a scattering wave for $l = 0$ and $20\,\text{MeV}$ in Fig. 13.11. At small distances our potential $V(r)$ is so deep that the term $Eu(r)$ in (13.3) is scarcely important compared with the term $V(r)u(r)$. For this reason the first two zeros of the wave function are fairly insensitive to a variation of the energy E. An equally appropriate explanation, which of course is linked with that just given, is that the scattering wave functions must be orthogonal to the bound state functions, and that this condition has a dominant influence on the form of the scattering wave functions at short distances.

b) Figure 13.12 shows a scattering wave function for $l = 4$, $E = 5\,\text{MeV}$, and Fig. 13.13 shows for comparison a scattering wave function with $l = 10$ and $E = 10\,\text{MeV}$. Notice the strong influence of the centrifugal barrier, in that the wave function with $l = 10$ is suppressed still more strongly in the neighbourhood of the origin than the wave function with $l = 4$.

Radial Wave Function (unnormalized)

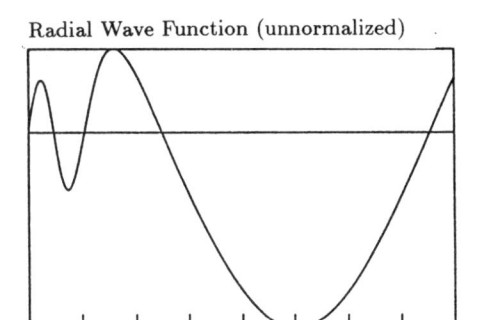

Fig. 13.10. Scattering wave function for $l = 0$ and $E = 1\,\text{MeV}$

Radial Wave Function (unnormalized)

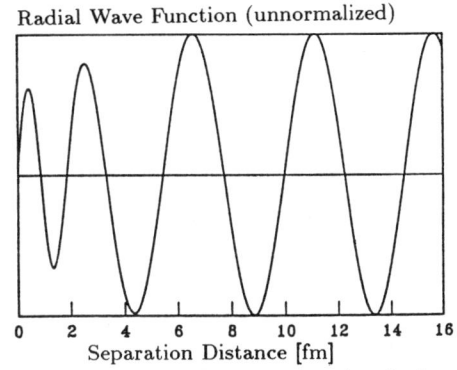

Fig. 13.11. Scattering wave function for $l = 0$ and $E = 20\,\text{MeV}$

Radial Wave Function (unnormalized)

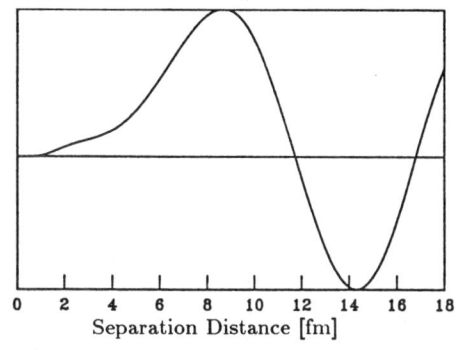

Fig. 13.12. Scattering wave function for $l = 4$ and $E = 5\,\text{MeV}$

Radial Wave Function (unnormalized)

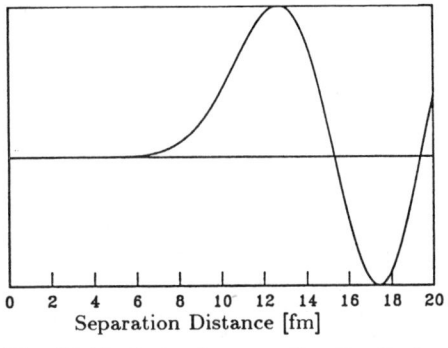

Fig. 13.13. Scattering wave function for $l = 10$ and $E = 10\,\text{MeV}$

From analytical investigations one knows that $u_l(r)$ behaves in the neighbourhood of the origin like r^{l+1}, i.e. $u_4(r)$ goes to zero like r^5 and $u_{10}(r)$ like r^{11}. The numerical display of the wave functions confirms this.

13.5.4 a) In order to incorporate the Coulomb potential into the program, one only needs to follow the IF-block in the FUNCTION subprogram V, i.e. line 208 of Fig. 13.4, with the

```
      DOUBLE PRECISION FUNCTION ERF(X)
      IMPLICIT DOUBLE PRECISION (A-H,O-Z)
      PARAMETER (A1=0.254829592D0, A2=-0.284496736D0)
      PARAMETER (A3=1.421413741D0, A4=-1.453152027D0)
      PARAMETER (A5=1.061405429D0, P=0.3275911D0)
      Y=ABS(X)
      T=1.D0/(1.D0+P*Y)
      IF(Y.LE.25.D0) THEN
        ERF=1.D0-(A1*T+A2*T*T+A3*T**3+A4*T**4+A5*T**5)*EXP(-Y*Y)
      ELSE
        ERF=1.D0
      ENDIF
      IF(X.LT.0.D0) ERF=-ERF
      END
```

Fig. 13.14. FUNCTION subprogram ERF

instruction

```
V=V+4.D0*1.2D0*1.2D0/R*ERF(0.75D0*R).
```

For the Gaussian error function erf(x) a further FUNCTION subprogram ERF must be made available. A suggested program for our proposed approximation is to be found in Fig. 13.14.

b) Because the resonance is so sharp we have to strain both the method and also the personal computer to the limits of their capabilities. We choose $i_b = 10\,000$, $b = 160$ fm, $l = 0$ and compute the wave functions once with $E = 0.091$ MeV and once with E=0.093 MeV. We see that the first major half-wave is positive in the first case and negative in the second case (see Fig. 13.15 and 13.16). Now we look, as in the case of bound states, for the energy at which the wave function changes sign, i.e. we halve the energy interval, compute, halve again, compute, and so on. The resonance state lies at $E = 0.091972$ MeV (the last decimal place may turn out somewhat different for computers with different rounding, but this is not important). Here the wave function has the shape shown in Fig. 13.17.

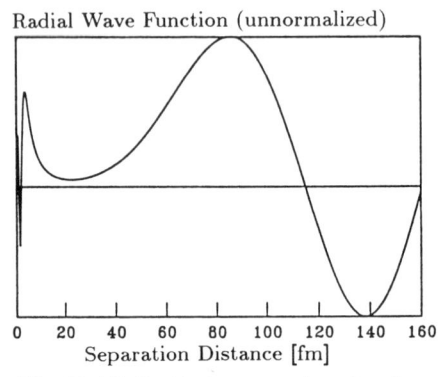

Fig. 13.15. Scattering wave function for $l = 0$ and $E = 0.091$ MeV

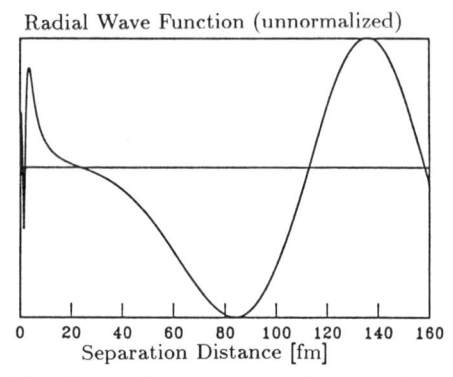

Fig. 13.16. Scattering wave function for $l = 0$ and $E = 0.093$ MeV

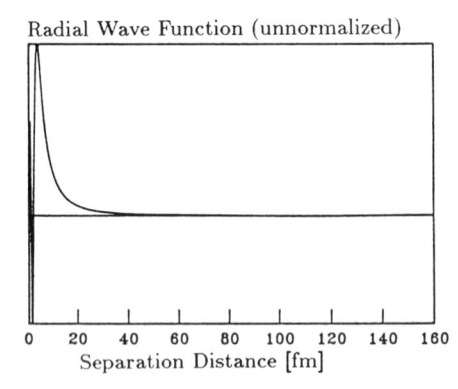

Fig. 13.17. The resonance wave function for $l = 0$ and $E = 0.091972$ MeV

A glance at the numerical display shows that, in the case of Fig. 13.15, the amplitude of the wave function in the inner space is about half as great as in the outer space, whereas in the case of Fig. 13.17 it is 250 times greater in the inner space than in the outer space. The probability of the two α-particles being close together is large at the energy of the resonance and small at all other energies. This becomes strongly apparent in the scattering cross-section.

14. The Quantum Mechanical Harmonic Oscillator

14.1 Formulation of the Problem

In Chap. 5 we have treated one-dimensional harmonic and anharmonic oscillations. The harmonic oscillation was characterised by the fact that the oscillation frequency did not depend on the amplitude of the displacement. If we had considered three-dimensional motion we should have found that for the harmonic oscillator all circular and elliptical orbits have the same frequency. In this sense the harmonic oscillation is the simplest of all periodic motions.

In this chapter we shall now consider the three-dimensional quantum mechanical harmonic oscillator. We shall seek stationary solutions, i.e. eigen-solutions of the time-independent Schrödinger equation.

The quantum mechanical harmonic oscillator also displays a special simplicity. It is one of the few systems for which the Schrödinger equation can be solved analytically. The eigenvalues are equally spaced. The eigen-states form a complete basis and are frequently used for the representation of operators. We shall go into this in more detail in the following chapters.

In the present chapter we shall be interested in the calculation of wave functions and in the connection between the quantum mechanical states and the classical motions. In the states with low excitation energies such a connection cannot be recognised because of the Heisenberg uncertainty principle. Only for high quantum numbers must a connection exist because of the correspondence principle. Since we do not make time-dependent calculations we shall not be able to track any particle paths. We shall, however, inspect the probability densities. For the classical orbit, the probability with which the moving body is to be found at any particular element of its orbit is inversely proportional to the velocity with which the body traverses that element of the orbit. This velocity, and hence the probability of location on the element of the orbit, can be readily calculated by classical means from the total energy and the potential energy at the element under consideration. It will be interesting to see how closely the quantum mechanical probability density reproduces the classical solution for high quantum numbers.

The spherically symmetric oscillator potential is

$$V(r) = cr^2 = \tfrac{1}{2}m\omega^2 r^2 \ . \qquad (14.1)$$

In the second part of the formula we have expressed the constant c by the mass m of the moving body and the angular frequency ω of the corresponding classical motion. With the oscillator potential the radial Schrödinger equation becomes (see (13.3))

$$-\frac{\hbar^2}{2m}\frac{d^2}{dr^2}v_l(r) + \frac{\hbar^2}{2m}\frac{l(l+1)}{r^2}v_l(r) + \tfrac{1}{2}m\omega^2 r^2 v_l(r) = E v_l(r) \quad . \tag{14.2}$$

We shall consider the eigen-solutions of this in the following section.

14.2 Numerical Method

We introduce as abbreviations the oscillator width parameter

$$a = \frac{m\omega}{2\hbar} \tag{14.3}$$

and the length

$$r_0 = \frac{1}{(2a)^{1/2}} = \left(\frac{\hbar}{m\omega}\right)^{1/2} \quad . \tag{14.4}$$

Equation (14.2) can be rearranged in the form

$$v_l''(r) - \left(\frac{r^2}{r_0^4} + \frac{l(l+1)}{r^2}\right)v_l(r) = -\frac{2mE}{\hbar^2}v_l(r) \quad . \tag{14.5}$$

The eigen-solutions of this equation are known analytically. We omit the derivation and for this refer to the literature [14.1]. The energy eigenvalues E of (14.5) are given by

$$E_{nl} = (2n + l - 1/2)\hbar\omega \quad . \tag{14.6}$$

The quantum number n is called the radial quantum number and takes the values $n = 1, 2, 3, \ldots$ The associated eigen-functions $v_{nl}(r)$ are called radial oscillator functions. They satisfy the following conditions

$$v_{nl}(0) = 0 \quad , \tag{14.7}$$

$$\int_0^\infty |v_{nl}(r)|^2 dr = 1 \tag{14.8}$$

and have the form

$$v_{nl}(r) = \left(\frac{2^{n+l+1}(n-1)!}{(2n+2l-1)!! r_0 \sqrt{\pi}}\right)^{1/2} \left(\frac{r}{r_0}\right)^{l+1} \mathcal{L}_{n-1}^{l+1/2}\left(\frac{r^2}{r_0^2}\right) \exp\left(\frac{-r^2}{2r_0^2}\right) \quad . \tag{14.9}$$

The functions

$$\mathcal{L}_{n-1}^{l+1/2}(x) = \sum_{k=0}^{n-1}(-1)^k \frac{(l+1/2)(l+3/2)\ldots(l-1/2+n)}{(l+1/2)(l+3/2)\ldots(l-1/2+k)} \frac{x^k}{k!(n-1-k)!}$$

$$= \sum_{k=0}^{n-1} \frac{\Gamma(l+1/2+n)}{\Gamma(l+3/2+k)} \frac{(-x)^k}{k!(n-1-k)!} \tag{14.10}$$

are called generalised Laguerre polynomials. The lowest generalised Laguerre polynomials are

$$\mathcal{L}_0^{l+1/2}(x) = 1 \tag{14.11a}$$

and

$$\mathcal{L}_1^{l+1/2}(x) = l + 3/2 - x \ . \tag{14.11b}$$

The calculation of higher polynomials is most simply carried out by the recursion formula

$$\mathcal{L}_n^{l+1/2}(x) = \frac{1}{n}\left((-x + l - 1/2 + 2n)\mathcal{L}_{n-1}^{l+1/2}(x) - (n + l - 1/2)\mathcal{L}_{n-2}^{l+1/2}(x)\right) \ . \tag{14.12}$$

From (14.9) and (14.11) the radial oscillator functions for $n = 1$ and $n = 2$ are

$$v_{1l}(r) = \left(\frac{4 \cdot 2^l}{(2l+1)!! r_0 \sqrt{\pi}}\right)^{1/2} \left(\frac{r}{r_0}\right)^{l+1} \exp\left(\frac{-r^2}{2r_0^2}\right) \ , \tag{14.13a}$$

$$v_{2l}(r) = \left(\frac{8 \cdot 2^l}{(2l+3)!! r_0 \sqrt{\pi}}\right)^{1/2} \left(\frac{r}{r_0}\right)^{l+1} \left(l + 3/2 - \frac{r^2}{r_0^2}\right) \exp\left(\frac{-r^2}{2r_0^2}\right) \ . \tag{14.13b}$$

For the higher radial oscillator functions one obtains from (14.9) and (14.12) the recursion formula

$$v_{n+1,l}(r) = \left(\frac{1}{n(n+l+1/2)}\right)^{1/2}$$
$$\times \left(\left(-\frac{r^2}{r_0^2} + l - 1/2 + 2n\right) v_{nl}(r) - ((n+l-1/2)(n-1))^{1/2} v_{n-1,l}(r)\right) \ . \tag{14.14}$$

With this formula we shall calculate the functions $v_{nl}(r)$. As in Chap. 13 we choose an interval $[0, b]$ for the distance r, and divide it into i_b sub-intervals. The oscillator functions will then be computed at the mesh points $r_i = ih$.

Let us look once again at the radial Schrödinger equation (14.2). For the angular momentum quantum number $l = 0$ the centrifugal potential vanishes, and the oscillator potential exerts on the mass m a force towards the origin proportional to the displacement. The corresponding classical motion is the linear oscillation.

For $l \neq 0$ the centifugal potential appears in (14.2). For small values of r the centifugal potential is dominant, for larger values of r the oscillator potential is dominant. There is a potential well in the range of r, in which the mass moves. The classical path with non-vanishing angular momentum is the circular or elliptical orbit. If we ignore the angular motion and consider the classical path only with respect to the distance r from the origin, then the body moves to and fro between a minimum distance r_1 and a maximum distance r_2 (see Fig. 14.1). In the circular orbit $r_1 = r_2$.

$V_{\text{class}}(r)$

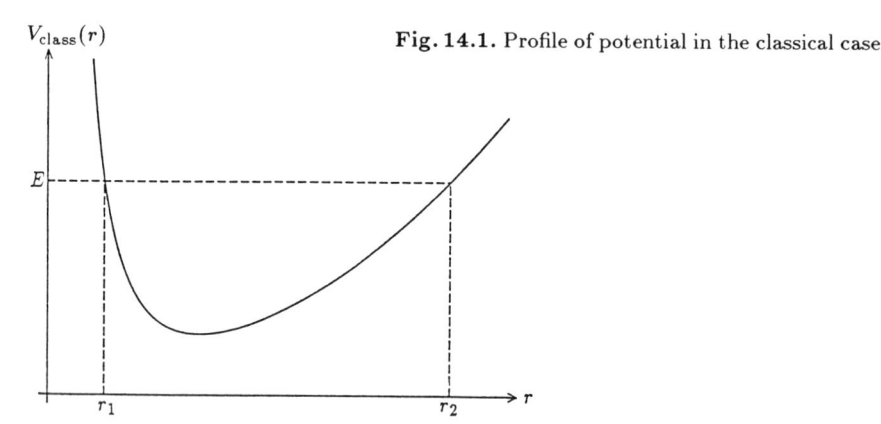

Fig. 14.1. Profile of potential in the classical case

r_1 $\quad r_2$

In place of the centifugal potential $\hbar^2 l(l+1)/2mr^2$ we have in the classical case the kinetic energy of the angular motion, i.e. the expression $L^2/2mr^2$, where L is the angular momentum. For the radial motion we then obtain the "potential"

$$V_{\text{class}}(r) = \frac{1}{2}m\omega^2 r^2 + \frac{L^2}{2mr^2} \quad . \tag{14.15}$$

For the reversal points we have

$$E = V_{\text{class}}(r_{1,2}) \quad , \tag{14.16}$$

from which it follows that

$$r_{1,2} = \left(\frac{E \pm (E^2 - \omega^2 L^2)^{1/2}}{m\omega^2}\right)^{1/2} \quad . \tag{14.17}$$

From the radial portion of the kinetic energy

$$T = E - V_{\text{class}}(r) = \tfrac{1}{2}mv^2 \tag{14.18}$$

it follows that the radial velocity is

$$v = \left(\frac{2E}{m} - \omega^2 r^2 - \frac{L^2}{m^2 r^2}\right)^{1/2} \quad . \tag{14.19}$$

One thus obtains for the classical probability density $w(r)$ at a point r in the interval (r_1, r_2) the formula:

$$w(r) = \frac{2\omega}{\pi}\frac{1}{|v|} = \frac{1}{\frac{\pi}{2}[2E/(m\omega^2) - r^2 - L^2/(m^2\omega^2 r^2)]^{1/2}} \quad . \tag{14.20}$$

This formula is to be compared with the quantum mechanical probability density

$$w_{nl}(r) = |v_{nl}(r)|^2 \quad . \tag{14.21}$$

We substitute for E the quantum mechanical energy eigenvalue (14.6) and for L^2 the eigenvalue

$$L^2 = \hbar^2 l(l+1) \quad . \tag{14.22}$$

We accordingly obtain from (14.20), together with (14.4),

$$w(r) = \frac{1}{\frac{\pi}{2}[(4n+2l-1)r_0^2 - r^2 - l(l+1)r_0^4/r^2]^{1/2}} \tag{14.23}$$

for $r_1 < r < r_2$.

14.3 Programming

In order to simplify the programming we introduce further abbreviations. The factor in front of the exponential function in (14.13a) will be denoted by $\sqrt{c_l}$, thus giving

$$v_{1l}(r) = \sqrt{c_l} \exp\left(\frac{-r^2}{2r_0^2}\right) \quad . \tag{14.24}$$

The quantities c_l can be computed recursively:

$$c_0 = \frac{4}{r_0 \sqrt{\pi}} \frac{r^2}{r_0^2} \quad , \tag{14.25}$$

$$c_l = \left(\frac{2}{2l+1} \frac{r^2}{r_0^2}\right) c_{l-1} \quad , \quad l \geq 1 \quad . \tag{14.26}$$

Equation (14.13b) can be expressed, using (14.13a), in the form

$$v_{2l}(r) = \left(\frac{2}{2l+3}\right)^{1/2} v_{1l}(r) \left(l + 3/2 - \frac{r^2}{r_0^2}\right) \quad . \tag{14.27}$$

The computation of the oscillator functions $v_{nl}(r)$ is carried out in subroutine RADOS (Fig. 14.3). The calling program, i.e. in our case the main program KAP14, transfers to RADOS the values for the distance r, the oscillator width parameter a, the angular momentum quantum number l and the maximal radial quantum number $n_{\max}$. The subroutine RADOS computes for all radial quantum numbers n from 1 to $n_{\max}$ the function values $v_{nl}(r)$ and transfers these

Physical title	FORTRAN title	Physical title	FORTRAN title
r	R	n	N
r_0	RO	c_l	CL
a	A	x	X
l	L	v_{nl}	OS(N)
$n_{\max}$	NMAX	π	PI

Fig. 14.2. Notation employed in subroutine RADOS

```
            SUBROUTINE RADOS(R,A,L,NMAX,OS)                              100

            IMPLICIT DOUBLE PRECISION (A-H,O-Z)                          101
            PARAMETER (PI=3.1415926535D0)                                102
            PARAMETER (EXPMAX=700.D0)                                    103
            DIMENSION OS(NMAX)                                           104
            IF((R.LT.0.D0).OR.(L.LT.0).OR.(NMAX.LT.1)) STOP              105
           &'Error in SUBROUTINE RADOS: False value for R, L or NMAX !'  106
            R0=1.D0/SQRT(2.D0*A)                                         107
            X=(R*R)/(R0*R0)                                              108
            CL=4.D0/(SQRT(PI)*R0)*X                                      109
            IF(L.GE.1) THEN                                              110
              DO 10 J=1,L                                                111
                CL=2*CL/(2*J+1)*X                                        112
    10      CONTINUE                                                     113
            ENDIF                                                        114
            IF(X/2.LE.EXPMAX) THEN                                       115
              OS(1)=SQRT(CL)*EXP(-X/2)                                   116
            ELSE                                                         117
              OS(1)=0.D0                                                 118
            ENDIF                                                        119
            IF(NMAX.GE.2) OS(2)=SQRT(2.D0/(2*L+3))*OS(1)*(L+1.5D0-X)     120
            IF(NMAX.GE.3) THEN                                           121
              DO 20 N=2,NMAX-1                                           122
                OS(N+1)=SQRT(1.D0/(N*(N+L+0.5D0)))*                      123
           &      ((2*N+L-0.5D0-X)*OS(N)-SQRT((N-1)*(N+L-0.5D0))*OS(N-1))124
    20      CONTINUE                                                     125
            ENDIF                                                        126
            END                                                          127
```

Fig. 14.3. Subroutine RADOS

as array OS to the calling program (see the parameter list in line 100). If RADOS is called with non-meaningful parameter values ($r < 0$, $l < 0$ or $n_{\max} < 1$), the program stops (lines 105 and 106). With the help of the notation listed in Fig. 14.2 and the relation

$$x = \frac{r^2}{r_0^2} \qquad (14.28)$$

the program instructions of RADOS can easily be understood. It suffices to list, item by item, which equation is treated where: (14.4) in line 107, (14.28) in line 108, (14.25) in line 109, (14.26) in the DO-loop from line 111, (14.24) in line 116, (14.27) in line 120 and finally (14.14) in the DO-loop from line 122. The IF-block of lines 115 to 119 is to prevent a possible exponent overflow.

We shall compute the oscillator functions on the interval $[0, b]$ with $b = 12\,\mathrm{fm}$. We accordingly divide the interval into $i_b = 480$ sub-intervals and obtain with the mesh width

$$h = \frac{b}{i_b} \qquad (14.29)$$

the mesh points

$$r_i = ih \quad , \quad i = 0, 1, 2, \ldots, i_b \quad . \qquad (14.30)$$

Physical title	FORTRAN title	Physical title	FORTRAN title
a	A	i_b	IB
l	L	h	H
$n_{\max}$	NMAX	r_i	R(I)
n	N	$v_{nl}(r_i)$	VNL(I), OS(N)
b	B	$w_{nl}(r_i)$	WNL(I)

Fig. 14.4. Notation employed in main program KAP14

```
        PROGRAM KAP14                                         200

        IMPLICIT DOUBLE PRECISION (A-H,O-Z)                   201
        PARAMETER (IB=480, NMAX=20, B=12.D0.)                 202
        DIMENSION R(0:IB), VNL(0:IB,NMAX), WNL(0:IB,NMAX), OS(NMAX)  203
*       (Input: A,L)

        H=B/IB                                                204
        DO 20 I=0,IB                                          205
        R(I)=I*H                                              206
        CALL RADOS(R(I),A,L,NMAX,OS)                          207
        DO 10 NN=1,NMAX                                       208
        VNL(I,NN)=OS(NN)                                      209
        WNL(I,NN)=VNL(I,NN)*VNL(I,NN)                         210
10      CONTINUE                                              211
20      CONTINUE                                              212

*       (Input: N)
*       (Output: VNL,WNL)

        END                                                   213
```

Fig. 14.5. Numerical portion of main program KAP14

We now consider the main program KAP14 (Fig. 14.5). In lines 204 and 206 we recognise again the last two equations, where the computation of the mesh points is embedded in a DO-loop which forms almost the entire numerical portion (lines 205 to 212). After the computation of a grid point the subroutine RADOS is called (line 207). The main program receives in the one-dimensional array OS the function values $v_{nl}(r_i)$ for the radial quantum numbers n from 1 to $n_{\max}$, with fixed l and r_i. For the output we store away, in the DO-loop of lines 208 to 211, the two-dimensional array VNL, containing the function values $v_{nl}(r_i)$ for all n with $1 \leq n \leq n_{\max}$ and all r_i with $i = 0, 1, \ldots, i_b$. Finally in line 210 the probability density $w_{nl}(r_i)$ is computed from (14.21).

The values for the number of sub-intervals i_b, for the maximal principal quantum number $n_{\max}$ and for the upper limit b of the distance interval are specified in a PARAMETER instruction (line 202). The oscillator width parameter a, the angular momentum quantum number l and the radial quantum number n are input. The functions $v_{nl}(r)$ and $w_{nl}(r)$ are output in graphical and, optionally, in numerical form. The program here summarised is to be found on the diskette in file KAP-14.FOR.

14.4 Exercises

14.4.1 Consider the radial oscillator functions $v_{nl}(r)$ and the associated probability densities $w_{nl}(r)$ for various quantum numbers n and l and for various oscillator width parameters a. Notice the differences between states with $l = 0$ and those with $l \neq 0$. Look at probability densities for high quantum numbers n. Can one recognise a qualitative connection between the quantum mechanical probability density at high quantum numbers and the classical probability density?

14.4.2 Write a program which computes the classical probability density $w(r)$ from (14.23) and displays it graphically. Compare these densities $w(r)$ with the quantum mechanical densities $w_{nl}(r)$.

14.5 Solutions to the Exercises

14.5.1 The oscillator functions show $n - 1$ zeros. They also vanish as $r \to 0$ and as $r \to \infty$.

The classical motion has a reversal at r_2 (see Fig. 14.1). The region $r > r_2$ is forbidden in the classical motion. In quantum mechanics there is a tunnel effect, which allows a particle to penetrate into the forbidden region. The nature of the penetration into the "tunnel" is described by the form of the wave function. The latter is determined by the potential and the energy, i.e. by the form and height of the "hill" over the tunnel. For large r the hill over the tunnel increases essentially as r^2. The corresponding decay of the wave function is similar to the decay of a Gauss function at large r-values. For non-zero angular momentum the particle is kept at a distance from the origin by the angular momentum barrier. Towards the origin the wave function decreases like r^{l+1}. The different behaviour of the oscillator functions in the forbidden region can be well demonstrated in the graphical presentation by suitable choice of the oscillator width parameter a. Figure 14.6 shows the harmonic oscillator function and the square of its modulus for $n = 7$, $l = 5$ and $a = 0.5\,\text{fm}^{-2}$.

For large values of n the quantum mechanical probability density $w_{nl}(r)$ approaches the classical probability density $w(r)$. The interference pattern of the function $w_{nl}(r)$ is a typical quantum mechanical phenomenon and does not admit of a classical explanation. The

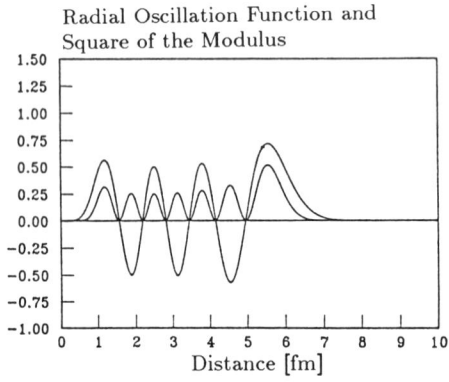

Fig. 14.6. Harmonic oscillator function $v_{nl}(r)$ and probability density $|v_{nl}(r)|^2$ for $n = 7$, $l = 5$ and $a = 0.5\,\text{fm}^{-2}$

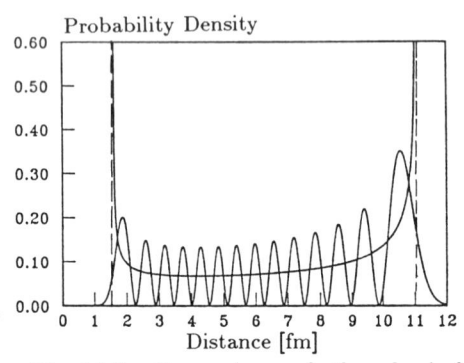

Fig. 14.7. Comparison of the classical probability density $w(r)$ with the quantum mechanical probability density $w_{nl}(r)$ for $n = 14$, $l = 10$ and $a = 0.3\,\text{fm}^{-2}$

envelope of the function $w_{nl}(r)$, however, already at $n = 7$ shows a similarity with the corresponding classical function $w(r)$.

14.5.2 In the computation of $w(r)$ from (14.23) it is to be noticed that $w(r)$ vanishes when the argument of the root is negative. The programming exercise is most easily carried out by calculating the function $w(r)$ in the main program between lines 210 and 211 and overwriting $v_{nl}(r)$ with $w(r)$. Then the graphical output produces the curves of $w(r)$ and $w_{nl}(r)$ in the same picture and makes an immediate comparison possible. Figure 14.7 shows such a comparison for $n = 14$, $l = 10$ and $a = 0.3\,\text{fm}^{-2}$.

15. Solution of the Schrödinger Equation in Harmonic Oscillator Representation

15.1 Formulation of the Problem

In Chaps. 13 and 14 we have become acquainted with the Schrödinger equation as a differential equation. It was established in this form by Erwin Schrödinger in 1926 [15.1]. Already a year earlier Werner Heisenberg had founded quantum mechanics in the form of a matrix equation [15.2]. It later became apparent that both equations represented the same physical theory in different mathematical formulations: the equations of Heisenberg and Schrödinger can be transformed into one another.

We shall carry out the transformation from the Schrödinger equation considered in Chap. 13 to the matrix equation, using the function space spanned by the harmonic oscillator functions as the representation space. The Heisenberg matrix equation also can be separated into equations for angular and radial motions, provided the potential V is rotationally symmetric. We shall again assume that this separation has already been carried out, and concentrate on the treatment of the radial equation (13.3).

Formally one can write (13.3) as an operator equation,

$$H_l u_l = E u_l \quad . \tag{15.1}$$

Here H_l is the radial Hamilton operator and u_l the radial part of a quantum mechanical state, each relating to the angular momentum quantum number l. Both H_l and u_l are abstract quantities which acquire a numerical significance when represented in a space of basis states. In configuration space, H_l is the linear differential operator occurring in (13.3)

$$H_l = -\frac{\hbar^2}{2m}\frac{d^2}{dr^2} + \frac{\hbar^2}{2m}\frac{l(l+1)}{r^2} + V(r) \quad . \tag{15.2}$$

We can also use a complete function space as representation basis, such as the space of radial oscillator functions $v_{nl}(r)$ considered in the last chapter. Equation (15.1) then becomes the linear homogeneous system of equations

$$\sum_{n'=1}^{D} H^l_{nn'} u^l_{n'} = E u^l_n \quad , \quad n = 1, \ldots, D \quad , \quad D \to \infty \quad . \tag{15.3}$$

Here we have written the angular momentum index l as a superscript in order to avoid having more than two indices in a row. The quantities H^l_{nn} are for

$n = n'$ the expectation values of the Hamilton operator H^l in the oscillator states v_{nl}, and for $n \neq n'$ they are the transition matrix elements, i.e.

$$H^l_{nn'} = \int_0^\infty v_{nl}(r)\left(-\frac{\hbar^2}{2m}\frac{d^2}{dr^2} + \frac{\hbar^2}{2m}\frac{l(l+1)}{r^2} + V(r)\right)v_{n'l}(r)dr \quad . \tag{15.4}$$

Instead of a wave function $u_l(r)$, a solution obtained from (15.3) is a vector (u^l_n), whose components are the amplitudes of the state u_l in the basis of oscillator functions. The transformation from the solution vector (u^l_n) to the wave function $u_l(r)$ is

$$u_l(r) = \sum_{n=1}^{D} u^l_n v_{nl}(r) \quad . \tag{15.5}$$

Equation (15.3) is an eigenvalue equation. It will therefore have solutions only for certain values of E. In the region of negative energies these are bound states. We shall calculate them for the potential used in Chap. 13, and compare them with the solutions obtained there.

In the region of positive energies (13.3) and (15.3) are not completely equivalent. In Chap. 13 we have seen that (13.3) has an asymptotically non-vanishing solution for any positive value of E. We can now investigate how the solutions of (15.3) look in the region of scattering energies. In practice one will always calculate with a finite number of basis states, i.e. with a finite dimension D of the matrix $(H^l_{nn'})$. There are accordingly only a finite number of eigenvalues. However, as the dimension increases, the eigenvalues in the region of positive energies will lie closer and closer together, and the eigenfunctions calculated from (15.5) will become more and more similar to the wave functions calculated from (13.3). The solutions obtained from the matrix equation will approach the solutions of the differential equation as far as this is possible with a basis of asymptotically vanishing functions. In fact, one can carry out scattering calculations with (15.3) also. One has to choose the dimension of the matrix $(H^l_{nn'})$ large enough, and one has to extend the eigenfunctions calculated from (15.3) and (15.5) into the asymptotic region.

15.2 Numerical Method

We calculate the matrix elements $H^l_{nn'}$ from (15.4). The contribution of the kinetic energy (including the centrifugal barrier) is a tri-diagonal matrix $(T^l_{nn'})$ with the analytically known elements [15.3]

$$T^l_{nn} = \left(2n + l - \frac{1}{2}\right)\frac{\hbar\omega}{2} \quad ,$$

$$T^l_{n,n+1} = T^l_{n+1,n} = \sqrt{n\left(n + l + \frac{1}{2}\right)}\frac{\hbar\omega}{2} \quad . \tag{15.6}$$

Accordingly we have

$$H^l_{nn'} = T^l_{nn'} + \int_0^\infty v_{nl}(r)V(r)v_{n'l}(r)dr \quad . \tag{15.7}$$

In the integral we replace the upper limit by a large value b and use the Simpson rule for the numerical evaluation.

If i_b is the number of integration steps, i.e. the number of meshes in the interval $[0, b]$, h the mesh width,

$$h = \frac{b}{i_b} \quad , \tag{15.8}$$

and $f(r)$ the integrand

$$f(r) = v_{nl}(r)V(r)v_{n'l}(r) \quad , \tag{15.9}$$

then (3.14) leads to the formula

$$H^l_{nn'} = T^l_{nn'} + \frac{h}{3}f(0) + \frac{4h}{3}f(h) + \frac{2h}{3}f(2h) + \ldots$$
$$\ldots + \frac{2h}{3}f(i_b h - 2h) + \frac{4h}{3}f(i_b h - h) + \frac{h}{3}f(i_b h) \quad . \tag{15.10}$$

The Simpson rule is especially simple in the present case. The first term with the factor $h/3$ vanishes, since $f(0) = 0$. The last terms of the sum vanishe, because in the neighbourhood of $r = b$ the potential $V(r)$ has become so small that $f(r)$ can be neglected. We therefore have only to evaluate a sum with the alternate weight factors $4h/3$ and $2h/3$, where it does not matter whether the number of terms is odd or even.

When we have determined the matrix elements $H^l_{nn'}$ from (15.10), we then have to solve the eigenvalue equation (15.3). The solution of the eigenvalue equation is equivalent to diagonalisation of the matrix $(H^l_{nn'})$. The eigenvectors $(u^l_n)_j$ are the column vectors of the unitary matrix which diagonalises $(H^l_{nn'})$. The eigenvalues E^l_j are the diagonal elements of the diagonalised matrix. Finally from (15.5) there follow the radial wave functions $u_{lj}(r)$ associated with the eigenvalues E^l_j.

The diagonalisation method which we shall employ is known by the name of the Jacobi rotation or the Jacobi method. We shall not describe it in detail, but refer the interested reader to the mathematical literature [15.4].

15.3 Programming

Since we consider the same example as in Chap. 13, we can fall back on the FUNCTION subprogram V used there, in order to compute the potential values $V(r_i)$ at the mesh points. We obtain the radial oscillator functions using the subroutine RADOS presented in Chap. 14.

In order to solve the eigenvalue problem we shall set up the subroutine EIGEN. Here we utilise an already existing ALGOL procedure [15.5], which we convert with suitable modifications into a FORTRAN program. EIGEN is called in the following way:

CALL EIGEN(HMAT, NMAX, NDIM, E, UMAT, H1, H2, H3)

The parameters HMAT, E, UMAT, H1, H2 and H3 are arrays which must be dimensioned in the calling program in the following way:

DOUBLE PRECISION HMAT(NMAX, NMAX), UMAT(NMAX, NMAX), H1(NMAX, NMAX)

DOUBLE PRECISION E(NMAX), H2(NMAX), H3(NMAX)

When EIGEN is called HMAT contains the matrix elements of the matrix to be diagonalised. It must be a real, symmetric matrix. NMAX specifies the dimension of the array HMAT, NDIM the dimension of the actual eigenvalue problem to be solved. In the case when NDIM is greater than NMAX, EIGEN breaks off with an error message. After EIGEN has been called E contains the eigenvalues of HMAT, and UMAT(N,J) the nth component of the eigenvector associated with the eigenvalue E(J). The eigenvectors are normalised, the eigenvalues arranged in ascending order. The auxiliary arrays H1, H2 and H3 only provide storage space needed for intermediate results. The array HMAT is not changed by EIGEN.

Figure 15.2 shows the main program KAP15. The notation used is to be found in Fig. 15.1. At first KAP15 prepares the mesh points r_i, the potential values $V(r_i)$ and the function values $v_{nl}(r_i)$ (lines 107–114). In line 115 we define the variable HQOM as abbreviation for the expression

$$\hbar\omega = 4a\frac{\hbar^2}{2m} \quad . \tag{15.11}$$

Physical title	FORTRAN title	Physical title	FORTRAN title
D	NDIM	$V(r_i)$	VR(I)
l	L	$f(r_i)$	F(I)
a	A	$v_{nl}(r_i)$	VNL(I,N), OS(N)
i_b	IB	$u_{lj}(r_i)$	U(I,J)
b	B	$\hbar^2/(2m)$	H2M
h	H	$(u_n^l)_j$	UMAT(N,J)
$\hbar\omega$	HQOM	E_j^l	E(J)
		$H_{nn'}^l$	HMAT(N,NP)
$r_i = i \cdot h$	R(I)	$V(r)$	V(R)

Fig. 15.1. Notation used in main program KAP15

```
              PROGRAM KAP15                                                 100

              IMPLICIT DOUBLE PRECISION (A-H,O-Z)                           101
              PARAMETER (IB=500, NMAX=20, B=10.D0, H2M=10.375D0)            102
              DIMENSION E(NMAX), H2(NMAX), H3(NMAX), OS(NMAX)               103
              DIMENSION HMAT(NMAX,NMAX), UMAT(NMAX,NMAX), H1(NMAX,NMAX)     104
              DIMENSION U(0:IB,NMAX), VNL(0:IB,NMAX), R(0:IB), VR(0:IB)     105

*      (Input: L,A,N)

              H=B/IB                                                        106
              DO 20 I=0,IB                                                  107
              R(I)=I*H                                                      108
              VR(I)=V(R(I))                                                 109
              CALL RADOS(R(I),A,L,NDIM,OS)                                  110
              DO 10 N=1,NDIM                                                111
              VNL(I,N)=OS(N)                                                112
10            CONTINUE                                                      113
20            CONTINUE                                                      114
              HQOM=4.D0*A*H2M                                               115
              DO 60 N=1,NDIM                                                116
              DO 50 NP=1,N                                                  117
               IF(NP.EQ.N) THEN                                             118
                 HMAT(N,NP)=(2.D0*N+L-0.5D0)*HQOM/2.D0                      119
               ELSE IF(NP.EQ.N-1) THEN                                      120
                 HMAT(N,NP)=SQRT(NP*(NP+L+0.5D0))*HQOM/2.D0                 121
               ELSE                                                         122
                 HMAT(N,NP)=0.D0                                            123
               ENDIF                                                        124
               DO 30 I=1,IB-1,2                                             125
                 F=VNL(I,N)*VR(I)*VNL(I,NP)                                 126
                 HMAT(N,NP)=HMAT(N,NP)+4.D0*H/3.D0*F                        127
30             CONTINUE                                                     128
               DO 40 I=2,IB-1,2                                             129
                 F=VNL(I,N)*VR(I)*VNL(I,NP)                                 130
                 HMAT(N,NP)=HMAT(N,NP)+2.D0*H/3.D0*F                        131
40             CONTINUE                                                     132
               HMAT(NP,N)=HMAT(N,NP)                                        133
50            CONTINUE                                                      134
60            CONTINUE                                                      135

*      (Output: HMAT)

              CALL EIGEN(HMAT,NMAX,NDIM,E,UMAT,H1,H2,H3)                    136
              DO 90 NP=1,NDIM                                               137
               DO 80 I=0,IB                                                 138
                 U(I,NP)=0.D0                                               139
                 DO 70 N=1,NDIM                                             140
                   U(I,NP)=U(I,NP)+UMAT(N,NP)*VNL(I,N)                      141
70               CONTINUE                                                   142
80             CONTINUE                                                     143
90            CONTINUE                                                      144

*      (Output: E,U)

              END                                                           145
```

Fig. 15.2. Numerical portion of main program KAP15

Then the matrix elements $H^l_{nn'}$ are evaluated from (15.10). The matrix elements $T^l_{nn'}$ are calculated from (15.6) (lines 118–124). In lines 125 to 132 one recognises the integration (15.10). In the calculation of $(H^l_{nn'})$ we make use of the fact that $(H^l_{nn'})$ is a symmetric matrix. We therefore calculate only the elements of the lower triangular matrix. By the instruction in line 133 we then obtain the full symmetric matrix $(H^l_{nn'})$.

In order to calculate the eigenvalues E^l_j of the matrix $(H^l_{nn'})$, the subroutine EIGEN is called in line 136. In the subsequent complex of DO-loops (lines 137–144) the function values $u_{lj}(r_i)$ of the wave functions are calculated from (15.5).

In the output the matrix elements $H^l_{nn'}$ are arranged in matrix form on the screen. For large D they can take up several screenfuls. The eigenvalues E appear in the form of a table. The wave function for a particular eigenvalue is output graphically and numerically. Whereas the eigenvalues are output in every case, all the following associated output is optional.

The program here summarised is to be found on the diskette in the file KAP-15.FOR.

15.4 Exercises

15.4.1 Study the matrix $(H^l_{nn'})$ for various oscillator width parameters a. Our suggestion for the range of variation of a is $0.1\,\text{fm}^{-2}$ to $1.0\,\text{fm}^{-2}$. It is possible to choose a so that the matrix $(H^l_{nn'})$ for $l = 0$ is almost diagonal in the first two oscillator states. When this is the case, do the values of the matrix elements H^0_{11} and H^0_{22} remind you of something?

15.4.2 Prepare the following diagram for each of $l = 0, 2, 4$: plot the oscillator width parameter a on the horizontal axis, and the energy eigenvalues E^l_j on the vertical axis as a function of a. Let a vary between $0.1\,\text{fm}^{-2}$ and $1.0\,\text{fm}^{-2}$. Plot the curves for various dimensions D of the basis of oscillator states. Take the values $D = 3, 5, 10, 15$. You then obtain a group of four curves for each energy eigenvalue E^l_j. What do you notice for (negative) binding energies, and what for (positive) scattering energies? How well do the binding energies agree with the values from Chap. 13?

15.4.3 Study the behaviour of the radial wave functions $u_{lj}(r)$ and compare the graphical representations with those obtained in Chap. 13. Choose a certain energy eigenvalue E^l_j, and find out for which values of a and D the wave function $u_l(r)$ obtained from Chap. 13 with $E = E^l_j$ can be approximated well. What is particularly noticeable for positive energies?

15.5 Solutions to the Exercises

15.5.1 Figure 15.3 shows the matrix $(H^l_{nn'})$ with $l = 0$, $D = 3$ for the oscillator width parameters $a = 0.2$ and $a = 0.65\,\text{fm}^{-2}$. In the second matrix the elements outside the diagonal are not large. On the diagonal one finds the elements -76.8 and $-26.8\,\text{MeV}$. The first of these values is a good approximation to the energy $-76.9\,\text{MeV}$ of the lowest bound state (see solution to Exercise 13.5.2). The second lies not far from the energy $-29.0\,\text{MeV}$ of the second bound state.

15.5.2 The larger the basis of harmonic oscillator states, the more closely are the true binding energies approximated by the eigenvalues of the Hamilton matrix, and the less sen-

-57.355769	-22.548768	-10.961965	-76.813707	-0.683996	-2.783760
-22.548678	-23.947651	-16.482265	-0.683996	-26.829173	6.691503
-10.961965	-16.482265	-6.741064	-2.783760	6.691503	14.741595
	$a = 0.2 \text{ fm}^{-2}$			$a = 0.65 \text{ fm}^{-2}$	

Fig. 15.3. The matrices $(H^l_{nn'})$ for $l = 0$, $D = 3$ and for specified values of the oscillator width parameter a

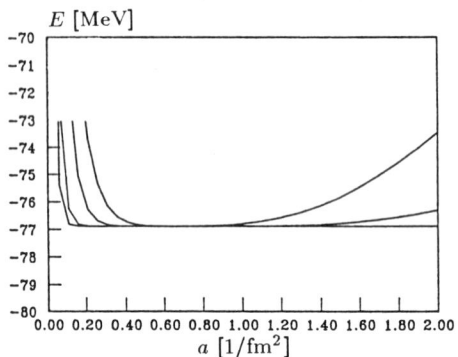

Fig. 15.4. Lowest energy eigenvalues for $l = 0$, $D = 3, 5, 10, 15$ as functions of the oscillator range parameter a

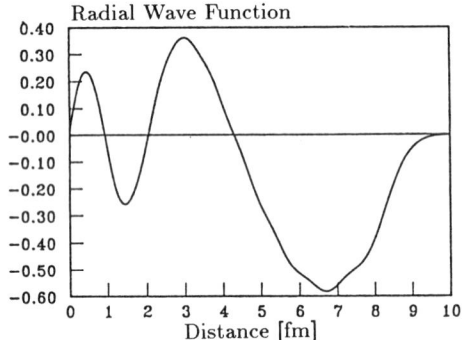

Fig. 15.5. Eigensolution of (15.3) for the positive eigenvalue $E_4 = 4.3471\,\text{MeV}$; the parameters are $l = 0$, $D = 20$, $a = 0.5\,\text{fm}^{-2}$. In the region $r > 5\,\text{fm}$ the agreement with the corresponding solution of the Schrödinger equation becomes poor

sitive is the approximation to variations in the range parameter a. Figure 15.4 shows, as a function of a, the lowest eigenvalues of the matrices $(H^l_{nn'})$ formed with 3, 5, 10 and 15 harmonic oscillator functions.

15.5.3 When the approximation to the true binding energy by an eigenvalue of (15.3) has an accuracy of several decimal places, then the true wave function is also approximated with satisfactory accuracy. For positive energies, i.e. in the region of scattering solutions, the wave functions of Chap. 13 are still approximated fairly well for small r, if the dimension D is large enough and if the range parameter a lies in the region in which the negative energy eigenvalues have their plateau (see Fig. 15.4). For larger values of r the approximation must break down, since the states of a finite basis have only a limited range. Figure 15.5 shows such a case.

16. The Ground State of the Helium Atom by the Hylleraas Method

16.1 Formulation of the Problem

In Chaps. 13 and 15 we have seen that the Schrödinger equation can be solved relatively simply for the motion of a particle in the spherically symmetric potential $V(r)$. The Schrödinger equation for the relative motion of two particles, which interact with each other by a spherically symmetric potential, can be reduced to this case. It is more difficult when the potential is no longer spherically symmetric or when three particles mutually interact.

An interesting three particle system is the helium atom. An α-particle with nuclear charge number $Z = 2$ forms the nucleus of the atom, around which two electrons move. The nucleus is about 7000 times heavier than an electron and may be considered to be at rest. The electrons move in the Coulomb field of the atomic nucleus and also interact with each other. The calculation of the energy of the ground state was one of the touchstones for quantum mechanics which had been established in the years 1925/26. The calculation was achieved by Hylleraas in 1929 [16.1]. In this chapter we shall retrace Hylleraas's historic calculation. The use of a computer will enable us to improve the approximation found by Hylleraas and in this way to test the accuracy of the historic calculation.

As Hamilton operator for the helium atom we shall use the approximation

$$H = -\frac{\hbar^2}{2m}(\Delta_1 + \Delta_2) - \frac{Ze^2}{|r_1|} - \frac{Ze^2}{|r_2|} + \frac{e^2}{|r_1 - r_2|} \quad , \tag{16.1}$$

i.e. we neglect all magnetic and relativistic effects together with the motion of the atomic nucleus. The vectors r_1 and r_2 denote the positions of the two electrons, m the mass of an electron, e the elementary charge and Z the nuclear charge number. We wish to calculate the energy of the ground state and the associated space part of the wave function. The space wave function of the helium atom in the ground state is symmetric under an interchange of the two electrons. The Pauli principle is taken care of by an antisymmetric spin state.

We cannot solve the Schrödinger equation with the Hamilton operator (16.1) either analytically or by direct numerical means. For the numerical solution difficulties arise, not only from the two Laplace operators Δ_1 and Δ_2, but also from the multi-dimensional boundary condition which the wave function must satisfy. Going over to the matrix equation here offers a distinct advantage. The boundary condition can be satisfied from the outset by the choice of the basis functions. If one chooses as the basis a product of single particle states,

$$\Phi_{ij}(\mathbf{r}_1, \mathbf{r}_2) = \varphi_i(\mathbf{r}_1)\varphi_j(\mathbf{r}_2) \quad , \tag{16.2}$$

then the Laplace operators also present no difficulty. The only difficulty arises from the (perhaps high) dimension of the matrix equation, because the dimension of a product basis is the product of the dimensions of the bases. If we were, for example, to use product states of the harmonic oscillator as basis, then we would need very many states and would quickly overwhelm our computer.

Hylleraas found a way out [16.1]. He constructed a relatively small number of basis states which are suited to the physical problem. The basis states contain a non-linear parameter α which may be freely varied.

As described in the previous chapter, the matrix equation

$$\sum_{n'=1}^{D} H_{nn'} a_{n'} = E a_n \quad , \quad n = 1, \ldots, D \quad , \tag{16.3}$$

is formulated with the Hamilton operator H and the chosen basis of dimension D. The matrix elements of the Hamilton operator depend on the variation parameter α. Accordingly all the eigenvalues E_j and all the associated eigenvectors $(a_n)_j$ depend on the parameter α.

One now applies the theorem of Hylleraas and Undheim [16.2]. This states that the eigenvalues E_j of (16.3) cannot be lower than the exact eigenvalues of the Schrödinger equation, regardless of special properties of the chosen function basis, and regardless also of the dimension D and of the variation parameter α. It is therefore always true that

$$E_{i,\text{exact}} \leq E_i \quad , \tag{16.4}$$

provided that the eigenvalues are arranged in ascending order. The eigenvalues of the matrix equation (16.3) are therefore upper bounds for the respective true eigenvalues. The theorem holds in the region of the bound states; it is trivially true in the region of scattering states, since in the immediate neighbourhood of the threshold $E = 0$ there is an infinite number of true states.

The theorem of Hylleraas and Undheim is used, not only to find a best approximation for the true eigenvalues, but also to assess the quality of the approximation:

1. One carries out the calculation of the eigenvalues from (16.3) for various values of the parameter α and obtains the minimum for the eigenvalue in which one is interested. The minimum is the desired approximation for the eigenvalue.
2. One plots $E_j(\alpha, D)$ as a function of α for various values of D and assesses the quality of the approximation from the plot (see Fig. 16.1). As long as the functions exhibit no plateau and as long as the minima of the functions do not converge with increasing values of D, one has not yet found a good approximation. The inverse of the statement is unfortunately not true; it can happen, for example, that all the basis states are orthogonal

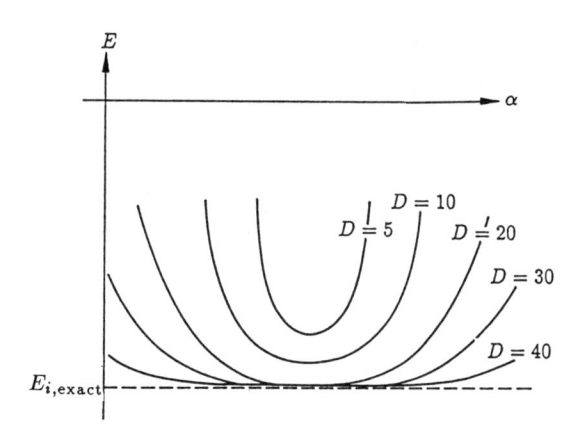

Fig. 16.1. Convergence behaviour of the ith eigenvalue in the Hylleraas method

to a component contained in the true solution and that therefore the true solution is not found, although the above criteria are fulfilled.

In the following section we shall consider the function basis introduced by Hylleraas for the ground state of the helium atom, and set up the eigenvalue equation.

16.2 Setting up the State Basis and the Matrix Equation

If there were no interaction between the electrons, one could determine the ground state wave function for the helium atom analytically. One would only have to put each electron into the ground state of the Schrödinger equation for the one-electron problem. One would have

$$\Phi_G(\boldsymbol{r}_1, \boldsymbol{r}_2) = \exp\left(\frac{-Z}{r_0}(r_1 + r_2)\right) \quad , \quad \text{where} \tag{16.5}$$

$$r_0 = \frac{\hbar^2}{me^2} = 0.52917 \, \text{Å} \tag{16.6}$$

is the Bohr radius. From now on we use the abbreviations r_1 and r_2 for $|\boldsymbol{r}_1|$ and $|\boldsymbol{r}_2|$. If one calculates the expectation value of the Hamilton operator (16.1) with the (normalized) state (16.5), this corresponds to a perturbation calculation of the first order for the electron-electron interaction. One certainly obtains an estimate of the binding energy, but it still differs from the true value by several eV.

The estimate can be improved by taking account of the fact that each electron shields the other electron from a portion of the nucleus-electron interaction. This is most easily achieved by replacing the nuclear charge number Z in (16.5) by an effective nuclear charge number $Z' = Z/\alpha$:

$$\Phi'_G(\boldsymbol{r}_1, \boldsymbol{r}_2) = \exp\left(\frac{-Z}{\alpha r_0}(r_1 + r_2)\right) \quad . \tag{16.7}$$

Since the theorem of Hylleraas and Undheim is valid also for $D = 1$ (the Ritz variational principle!) one may vary α and thus obtain a better estimate of the binding energy. The expression (16.7), however, does not take account of the fact that, because of their mutual repulsion, the electrons have a reduced probability of being close together, as this expression is purely a product of single particle wave functions. However, one can take (16.7) as the starting-point for the construction of a function basis, by multiplying Φ'_G by suitable factors. The following basis functions, obtained in this way, were introduced by Hylleraas:

$$\tilde{\Phi}_{jkm}(\mathbf{r}_1, \mathbf{r}_2)$$
$$= (r_1 + r_2)^j (r_1 - r_2)^k |\mathbf{r}_1 - \mathbf{r}_2|^m \Phi'_G(\mathbf{r}_1, \mathbf{r}_2) \quad (j, k, m \geq 0) \quad . \tag{16.8}$$

The tilde indicates that the basis functions are not orthogonal to one another. Only even values are allowed for k, because the space wave function in the ground state must be symmetric with regard to interchange of the two electrons. The wave function is written as

$$\psi(\mathbf{r}_1, \mathbf{r}_2) = \sum_{j,k,m} \tilde{a}_{jkm} \tilde{\Phi}_{jkm}(\mathbf{r}_1, \mathbf{r}_2) \quad . \tag{16.9}$$

The elements of the function basis with $m > 0$ enable us to allow approximately for the effect of the mutual repulsion of the electrons. In the exercises we shall see that an improvement of the estimate for the energy obtained with the simple expression (16.7) arises mainly from the elements of the function basis with $m > 0$ or $k > 0$, whereas the terms with $m = 0$, $k = 0$, $j > 0$ make little contribution. In his historic calculation Hylleraas used only three states, namely $(j, k, m) = (0, 0, 0)$, $(0,0,1)$ and $(0,2,0)$. In what follows we shall for the sake of brevity often denote the triad (j, k, m) by n.

The representation of the Schrödinger equation on the non-orthogonal basis (16.8) leads to the general eigenvalue equation

$$\sum_{n'} (\tilde{H}_{nn'} - E N_{nn'}) \tilde{a}_{n'} = 0 \quad , \quad \text{with} \tag{16.10}$$

$$\tilde{H}_{nn'} = \langle \tilde{\Phi}_n | H | \tilde{\Phi}_{n'} \rangle \quad \text{and} \tag{16.11}$$

$$N_{nn'} = \langle \tilde{\Phi}_n | \tilde{\Phi}_{n'} \rangle \quad . \tag{16.12}$$

We have to transform the general eigenvalue equation (16.11) to the special form (16.3) in order to be able to apply the theorem of Hylleraas and Undheim. As we shall see, the transformation is equivalent to orthonormalising the function basis (16.8).

The matrix N occurring in (16.10) and (16.12) is called the norm matrix. It contains all the information on the non-orthogonality of the function basis (16.8). The matrix N is real and symmetric. Because of the linear independence of the basis it is also positive definite, i.e. all its eigenvalues are greater than zero. Also N^{-1}, $N^{1/2}$ and $N^{-1/2}$ therefore exist. With the matrix $N^{-1/2}$ we can orthonormalise the basis (16.8).

We set
$$\Phi_n = \sum_{n'} (N^{-1/2})_{nn'} \tilde{\Phi}_{n'} \quad , \tag{16.13}$$

where $(N^{-1/2})_{nn'}$ denotes the nn' matrix element of the matrix $N^{-1/2}$. The basis of the transformed states Φ_n is orthonormal,
$$\langle \Phi_n | \Phi_{n'} \rangle = \delta_{nn'} \quad . \tag{16.14}$$

For the matrix elements of H on the orthonormalised basis we obtain
$$H_{nn'} = \langle \Phi_n | H | \Phi_{n'} \rangle = (N^{-1/2} \tilde{H} N^{-1/2})_{nn'} \quad . \tag{16.15}$$

The last expression is again the nn' element of the matrix product in brackets. Finally we transform the coefficients $\tilde{a}_n$ occurring in (16.9) and (16.10). From
$$\psi = \sum_n \tilde{a}_n \tilde{\Phi}_n = \sum_n a_n \Phi_n \tag{16.16}$$

and the transformation (16.13) one obtains
$$\tilde{a}_n = \sum_{n'} (N^{-1/2})_{nn'} a_{n'} \quad , \quad a_n = \sum_{n'} (N^{1/2})_{nn'} \tilde{a}_{n'} \quad . \tag{16.17}$$

We can now see how the basis transformation (16.13) affects the general eigenvalue equation (16.10). In matrix notation (16.10) reads
$$(\tilde{H} - NE)\tilde{a} = 0 \quad , \tag{16.18}$$

where $\tilde{a}$ denotes the column vector formed with the components $\tilde{a}_n$. We insert the unit matrix in the form $\mathbf{1} = N^{-1/2} N^{1/2}$ in (16.18) between $(\tilde{H} - NE)$ and $\tilde{a}$, multiply by $N^{-1/2}$ from the left and obtain
$$(N^{-1/2} \tilde{H} N^{-1/2} - E N^{-1/2} N N^{-1/2}) N^{1/2} \tilde{a} = 0 \quad . \tag{16.19}$$

The expression $N^{-1/2} \tilde{H} N^{-1/2}$ is equal to the matrix $(H_{nn'})$ appearing in (16.15), the expression $N^{-1/2} N N^{-1/2}$ is the unit matrix, and $N^{1/2} \tilde{a}$ is according to (16.17) equal to the column vector a with the components a_n. We see that (16.19) is identical with the special eigenvalue equation (16.3).

Th orthonormalisation of the basis (16.8) has led from the general eigenvalue equation (16.10) via (16.19) to the special eigenvalue equation (16.3). The equations (16.3) and (16.10) differ only in the representation basis of the operators. Not only the eigenvalues but also the wave functions ψ calculated from the eigenvectors are the same for both equations. With (16.19) we have also obtained a scheme by which the general eigenvalue equation can be solved numerically. We shall therefore not explicitly carry out the orthonormalisation of the basis (16.8) in our program. Instead we shall set up the general eigenvalue equation (16.10) and write a subroutine for the solution which works essentially according to the scheme of (16.19). The subroutine will be independent of the particular example that we are considering.

Before we turn to the programming of the eigenvalue equation, we have still to define the matrix elements $N_{nn'}$ and $\tilde{H}_{nn'}$. A numerical integration would be very expensive, since one would have to integrate over three independent coordinates. Hylleraas calculated the matrix elements analytically and we shall follow his example. The analytical calculation is not complicated, but tedious. It will therefore be carried out only for the elements $N_{nn'}$. For $\tilde{H}_{nn'}$ we shall give the results without the calculation.

With $n = (j, k, m)$, $n' = (j', k', m')$ and

$$\lambda = Z/(\alpha r_0) \tag{16.20}$$

we obtain from (16.12) and (16.8) the equation

$$N_{nn'} = \langle \tilde{\Phi}_{jkm} | \tilde{\Phi}_{j'k'm'} \rangle$$

$$= \int d^3r_1 d^3r_2 \tilde{\Phi}_{jkm}(\mathbf{r}_1, \mathbf{r}_2) \tilde{\Phi}_{j'k'm'}(\mathbf{r}_1, \mathbf{r}_2)$$

$$= \int d^3r_1 d^3r_2 (r_1 + r_2)^{j+j'} (r_1 - r_2)^{k+k'} |\mathbf{r}_1 - \mathbf{r}_2|^{m+m'}$$

$$\times \exp(-2\lambda(r_1 + r_2)) \quad . \tag{16.21}$$

We notice that the exponents contain only the sums $j + j'$, $k + k'$ and $m + m'$, so we introduce the abbreviations

$$J = j + j' \quad , \quad K = k + k' \quad , \quad M = m + m' \quad . \tag{16.22}$$

The integrand depends only on r_1, r_2 and the angle γ between $\mathbf{r}_1$ and $\mathbf{r}_2$. The integration over the other angular coordinates gives a factor $8\pi^2$. We obtain

$$N_{nn'} \equiv N_{JKM} = 8\pi^2 \int_0^\infty dr_1 \int_0^\infty dr_2 \int_{-1}^1 d(\cos\gamma) \left(r_1^2 r_2^2 (r_1 + r_2)^J (r_1 - r_2)^K \right.$$

$$\left. \times (r_1^2 + r_2^2 - 2r_1 r_2 \cos\gamma)^{M/2} \exp(-2\lambda(r_1 + r_2)) \right) \quad . \tag{16.23}$$

With the substitutions

$$u = \cos\gamma \quad , \quad v = r_1 + r_2 \quad , \quad w = r_1 - r_2 \tag{16.24}$$

the integral (16.23) clearly becomes

$$N_{JKM} = 4\pi^2 \int_0^\infty dv \int_{-v}^v dw \int_{-1}^1 du$$

$$\left(\frac{v^2 - w^2}{4} \right)^2 v^J w^K \left(\frac{v^2 + w^2 - u(v^2 - w^2)}{2} \right)^{M/2} \exp(-2\lambda v) \quad . \tag{16.25}$$

Integration over u gives

$$N_{KJM} = \frac{\pi^2}{M+2} \int_0^\infty dv \int_{-v}^v dw (v^2 - w^2) v^J w^K (v^{M+2} - |w|^{M+2}) \exp(-2\lambda v) \quad .$$

$$\tag{16.26}$$

Now we carry out the integration over w. For odd values of K the integral vanishes, since the integrand is then an odd function of w. For even values of K we obtain

$$N_{KJM} = \frac{2\pi^2}{M+2} \int_0^\infty dv$$

$$\left(\frac{1}{K+1} - \frac{1}{K+3} - \frac{1}{K+M+3} + \frac{1}{K+M+5} \right) v^{J+K+M+5} \exp(-2\lambda v) \ .$$
(16.27)

Finally the integration over v gives

$$N_{nn'} \equiv N_{JKM}$$
$$= 2\pi^2 (J+K+M+5)! \left(\frac{1}{2\lambda}\right)^{J+K+M+6} \left(\frac{1}{M+2}\right)$$
$$\times \left(\frac{1}{K+1} - \frac{1}{K+3} - \frac{1}{K+M+3} + \frac{1}{K+M+5} \right)$$

for $K = 0, 2, 4, \ldots$. (16.28a)

We have already seen that

$$N_{nn'} \equiv N_{JKM} = 0 \quad \text{for} \quad K = 1, 3, 5, \ldots \ . \tag{16.28b}$$

The Hamilton operator (16.1) consists of three parts,

$$H = T + C + W \ , \tag{16.29}$$

with

$$C = -\frac{Ze^2}{r_1} - \frac{Ze^2}{r_2} \ , \quad T = -\frac{\hbar^2}{2m}(\Delta_1 + \Delta_2) \ , \quad W = \frac{e^2}{|r_1 - r_2|} \ .$$
(16.30)

Without going in detail into the analytical calculation, we give the formulae for the matrix elements. The matrix elements of the Coulomb potential between the atomic nucleus and the two electrons are

$$\tilde{C}_{nn'} = \langle \tilde{\Phi}_{jkm} | C | \tilde{\Phi}_{j'k'm'} \rangle = -Ze^2 \hat{C}_{JKM} \ , \tag{16.31}$$

with

$$\hat{C}_{JKM} = \begin{cases} 8\pi^2 (J+K+M+4)! \left(\frac{1}{2\lambda}\right)^{J+K+M+5} \left(\frac{1}{M+2}\right) \\ \quad \times \left(\frac{1}{K+1} - \frac{1}{K+M+3}\right) & \text{for } K = 0, 2, 4, \ldots, \\ 0 & \text{for } K = 1, 3, 5, \ldots \end{cases}$$
(16.32)

The matrix elements for the Coulomb interaction between the electrons are

$$\tilde{W}_{nn'} = \langle \tilde{\psi}_{jkm} | W | \tilde{\psi}_{j'k'm'} \rangle = e^2 \hat{W}_{JKM} \quad , \tag{16.33}$$

with

$$\hat{W}_{JKM} = N_{JK,M-1} \quad . \tag{16.34}$$

In order to determine the $\tilde{W}_{JK,0}$ we need $N_{JK,-1}$. The derivation (16.23) to (16.28) goes through also for $M = -1$, so that this presents no problem.

The matrix elements of the kinetic energy of the electrons are

$$\tilde{T}_{nn'} = \langle \tilde{\psi}_{jkm} | T | \tilde{\psi}_{j'k'm'} \rangle = \frac{\hbar^2}{2m} \hat{T}_{jkmj'k'm'} \quad , \tag{16.35}$$

with

$$\begin{aligned}
\hat{T}_{jkmj'k'm'} = &\; 2[\lambda^2 N_{JKM} - J\lambda N_{J-1,KM} \\
&+ jj' N_{J-2,KM} + kk' N_{J,K-2,M} + mm' N_{JK,M-2}] \\
&+ 1/2[- M\lambda(\hat{C}_{JKM} - \hat{C}_{J,K+2,M-2}) \\
&+ (mj' + m'j)(\hat{C}_{J-1,KM} - \hat{C}_{J-1,K+2,M-2}) \\
&+ (mk' + m'k)(\hat{C}_{J+1,K,M-2} - \hat{C}_{J-1,KM})] \quad . \tag{16.36}
\end{aligned}$$

In the case when M is equal to zero, the matrix elements $N_{JK,-2}$, $\tilde{C}_{J,K+2,-2}$, $\tilde{C}_{J-1,K+2,-2}$ and $\tilde{C}_{J+1,K,-2}$ in (16.36) have to be calculated. The expressions (16.28) and (16.32) are not defined in this case. However, the corresponding matrix elements in (16.36) are multiplied by zero. We can therefore nevertheless evaluate (16.36). In the numerical computation, however, we have to introduce special treatment for the case $M = -2$, in order to prevent the computer from trying to divide by zero.

As matrix elements of the Hamilton operator we now have

$$\tilde{H}_{nn'} = \tilde{T}_{nn'} + \tilde{C}_{nn'} + \tilde{W}_{nn'} \quad . \tag{16.37}$$

16.3 Programming

We first consider the eigenvalue equation (16.10), (16.18) in general, i.e. separately from the specific problem treated in this chapter. We start from the real symmetric matrix A and the real symmetric, positive definite matrix B and look for the eigenvalues α and the associated eigenvectors x of the general eigenvalue equation

$$Ax = \alpha Bx \quad . \tag{16.38}$$

We wish to bring this general eigenvalue equation into the form of the special eigenvalue equation, and for this purpose we shall first solve the eigenvalue equation

$$By = \beta y \quad . \tag{16.39}$$

We shall denote by (β) the diagonal matrix whose diagonal elements are the eigenvalues β, and by Y the orthogonal matrix whose columns are the normalised eigenvectors y. Then we have the relation

$$(\beta) = Y^T B Y \quad, \quad \text{or} \tag{16.40}$$

$$B = Y(\beta) Y^T \quad. \tag{16.41}$$

Since (β) is a diagonal matrix we can easily obtain $(\beta)^{1/2}$ and $(\beta)^{-1/2}$. To do this we have only to raise the diagonal elements to the corresponding powers. Now we multiply (16.38) from the left by $(\beta)^{-1/2} Y^T$ and insert between A and x, and also between B and x, the unit matrix in the from $\mathbf{1} = Y(\beta)^{-1/2}(\beta)^{1/2} Y^T$. Taking account of (16.41) we obtain the special eigenvalue equation

$$Cz = \alpha z \quad, \quad \text{with} \tag{16.42}$$

$$C = (\beta)^{-1/2} Y^T A Y (\beta)^{-1/2} \quad \text{and} \tag{16.43}$$

$$z = (\beta)^{1/2} Y^T x \quad. \tag{16.44}$$

The eigenvalues of (16.42) are identical with those of (16.38). We obtain the eigenvectors x of (16.38) from the eigenvectors z of (16.42) by means of the transformation

$$x = Y(\beta)^{-1/2} z \tag{16.45}$$

following from (16.44).

Unfortunately FORTRAN offers no matrix operations, so that we are compelled to write out the components. We use the symbols A_{ik}, B_{ik} and C_{ik} for the matrix elements of the matrices A, B and C, α_k and β_k for the eigenvalues and x_{ik}, y_{ik} and z_{ik} for the ith components of the eigenvectors corresponding to the kth eigenvalue (where the y_{ik} are equivalent to the matrix elements of Y). From (16.43) it then follows that

$$C_{ij} = \sum_{k=1}^{n} \beta_i^{-1/2} y_{ki} \left(\sum_{l=1}^{n} A_{kl} y_{lj} \right) \beta_j^{-1/2} \tag{16.46}$$

and from (16.45)

$$x_{ij} = \sum_{k=1}^{n} y_{ik} \beta_k^{-1/2} z_{kj} \quad. \tag{16.47}$$

For the solution of the two special eigenvalue equations (16.39) and (16.42) we already have available the subroutine EIGEN, with which we have become acquainted in Chap. 15. In order to solve the general eigenvalue equation (16.38) we set up the subroutine EIGENX, which, besides twice calling EIGEN, essentially has the task of working out the formulae (16.46) and (16.47). For a more detailed treatment of the eigenvalue problem of symmetric matrices we refer to [16.3].

Mathematical title	FORTRAN title	Mathematical title	FORTRAN title
A_{ik}	A(I,K)	n	N
B_{ik}	B(I,K)	x_{ik}	X(I,K)
C_{ik}	C(I,K)	y_{ik}	Y(I,K)
α_k	ALPHA(K)	z_{ik}	Z(I,K)
$\beta_k, \beta_k^{-1/2}$	BETA(K)		

Fig. 16.2. Notation used in subroutine EIGENX

```
          SUBROUTINE EIGENX(A,B,NMAX,N,ALPHA,X,Y,Z,C,BETA,H1,H2)        100
          IMPLICIT DOUBLE PRECISION (A-H,O-Z)                           101
          DIMENSION ALPHA(NMAX), BETA(NMAX), H1(NMAX), H2(NMAX)         102
          DIMENSION A(NMAX,NMAX), B(NMAX,NMAX), C(NMAX,NMAX)            103
          DIMENSION X(NMAX,NMAX), Y(NMAX,NMAX), Z(NMAX,NMAX)            104
          IF((N.LT.1).OR.(N.GT.NMAX)) STOP                              105
         &'Error in SUBROUTINE EIGENX: 1 ≤ N ≤ NMAX not satisfied'      106
          CALL EIGEN(B,NMAX,N,BETA,Y,X,H1,H2)                           107
          DO 10 I=1,N                                                   108
          IF(BETA(I).LE.0.D0) STOP 'Error in SUBROUTINE EIGENX: B       109
         & singular or not pos. definite'                               110
          BETA(I)=SQRT(1.D0/BETA(I))                                    111
10        CONTINUE                                                      112
          DO 40 K=1,N                                                   113
          DO 30 J=1,N                                                   114
          X(K,J)=0.D0                                                   115
          DO 20 L=1,N                                                   116
          X(K,J)=X(K,J)+A(K,L)*Y(L,J)                                   117
20        CONTINUE                                                      118
30        CONTINUE                                                      119
40        CONTINUE                                                      120
          DO 70 I=1,N                                                   121
          DO 60 J=1,N                                                   122
          C(I,J)=0.D0                                                   123
          DO 50 K=1,N                                                   124
          C(I,J)=C(I,J)+Y(K,I)*X(K,J)                                   125
50        CONTINUE                                                      126
          C(I,J)=BETA(I)*C(I,J)*BETA(J)                                 127
60        CONTINUE                                                      128
70        CONTINUE                                                      129
          CALL EIGEN(C,NMAX,N,ALPHA,Z,X,H1,H2)                          130
          DO 100 J=1,N                                                  131
          DO 90 I=1,N                                                   132
          X(I,J)=0.D0                                                   133
          DO 80 K=1,N                                                   134
          X(I,J)=X(I,J)+Y(I,K)*BETA(K)*Z(K,J)                           135
80        CONTINUE                                                      136
90        CONTINUE                                                      137
100       CONTINUE                                                      138
          END                                                           139
```

Fig. 16.3. Subroutine EIGENX

The parameter list of EIGENX (Fig. 16.3) is quite extensive (line 100): the calling program (in our case the main program KAP16) transfers to EIGENX the two-dimensional arrays A and B, which contain the elements of the matrices in question, the dimension N of the matrices together with the maximum permissible dimension NMAX. After successful completion EIGENX gives back to the calling program the one-dimensional array ALPHA with the eigenvalues α and the two-dimensional array X with the eigenvectors x. The other arrays serve as auxiliary arrays: apart from reserving storage space, nothing is transferred in these arrays. The program is interrupted if the condition $1 \leq N \leq NMAX$ is violated (line 106) or if the matrix B is not positive definite (line 110).

After the eigenvalue equation (16.39) has been solved by calling EIGEN in line 107, the quantities $\beta_k^{-1/2}$ are calculated in the following DO-loop (from line 108). In order to carry out the calculation of the matrix elements C_{ij} we first determine the expression enclosed in brackets in (16.46), since this does not depend on i (lines 113–120). As the array X is only required at the end of EIGENX, we can store the intermediate results in this. In the DO-loop complex from line 121 to line 128 the elements of the matrix C are now calculated. With the second call of EIGEN the eigenvalue equation (16.42) is solved (line 130). As with the first call of EIGEN, X is used as an auxiliary array. The formula (16.47) is evaluated in a further DO-loop complex (lines 131 to 138).

Before the call of EIGENX the matrix elements $H_{nn'}$ and $N_{nn'}$ of H and N have to be calculated. This calculation is mainly executed outside the main program KAP16 in the four FUNCTION subroutines NN, CHAT, THAT and WHAT (Fig. 16.5). We use NN to obtain N_{JKM} from (16.28), CHAT to obtain $\hat{C}_{JKM}$ from (16.32), THAT to obtain $\hat{T}_{jkmj'k'm'}$ from (16.36) and WHAT to obtain $\hat{W}_{JKM}$ from (16.34).

The transfer of λ is effected by the COMMON block /LAMCOM/ (lines 204, 217 and 232). The factorial FAK in NN and CHAT is calculated by the FUNCTION subprogram FAK (Fig. 16.5). We notice here that the parameter N is an INTEGER expression, whereas the result FAK(N) is of DOUBLE PRECISION type. This type declaration has been made, since the factorial very quickly ascends into the region of high values, which the computer cannot represent as INTEGER numbers.

Physical title	FORTRAN title	Physical title	FORTRAN title
j	J1	J	J
k	K1	K	K
m	M1	M	M
j'	J2	λ	LAMBDA
k'	K2	π	PI
m'	M2		

Fig. 16.4. Notation used in the FUNCTION subroutines NN, THAT, CHAT and WHAT

```fortran
      DOUBLE PRECISION FUNCTION NN(J,K,M)                               200
      IMPLICIT DOUBLE PRECISION (A-H,O-Z)                               201
      PARAMETER (PI=3.1415926535D0)                                     202
      DOUBLE PRECISION LAMBDA                                           203
      COMMON /LAMCOM/LAMBDA                                             204
      IF((K.NE.2*(K/2)).OR.(M.EQ.-2)) THEN                              205
       NN=0.D0                                                          206
      ELSE                                                              207
       NN=2.D0*PI*PI*FAK(J+K+M+5)*(0.5D0/LAMBDA)**(J+K+M+6)             208
     &      *1.D0/(M+2.D0)*(1.D0/(K+1.D0)-1.D0/(K+3.D0)                 209
     &      -1.D0/(K+M+3.D0)+1.D0/(K+M+5.D0))                           210
      ENDIF                                                             211
      END                                                               212

      DOUBLE PRECISION FUNCTION THAT(J1,K1,M1,J2,K2,M2)                 213
      IMPLICIT DOUBLE PRECISION (A-H,O-Z)                               214
      DOUBLE PRECISION NN                                               215
      DOUBLE PRECISION LAMBDA                                           216
      COMMON /LAMCOM/LAMBDA                                             217
      J=J1+J2                                                           218
      K=K1+K2                                                           219
      M=M1+M2                                                           220
      THAT=2.D0*(LAMBDA*LAMBDA*NN(J,K,M)-J*LAMBDA*NN(J-1,K,M)           221
     &      +J1*J2*NN(J-2,K,M)+K1*K2*NN(J,K-2,M)                        222
     &      +M1*M2*NN(J,K,M-2))+0.5D0                                   223
     &      *(-M*LAMBDA*(CHAT(J,K,M)-CHAT(J,K+2,M-2))                   224
     &      +(M1*J2+M2*J1)*(CHAT(J-1,K,M)-CHAT(J-1,K+2,M-2))            225
     &      +(M1*K2+M2*K1)*(CHAT(J+1,K,M-2)-CHAT(J-1,K,M)))             226
      END                                                               227

      DOUBLE PRECISION FUNCTION CHAT(J,K,M)                             228
      IMPLICIT DOUBLE PRECISION (A-H,O-Z)                               229
      PARAMETER (PI=3.1415926535D0)                                     230
      DOUBLE PRECISION LAMBDA                                           231
      COMMON /LAMCOM/LAMBDA                                             232
      IF((K.NE.2*(K/2)).OR.(M.EQ.-2)) THEN                              233
       CHAT=0.D0                                                        234
      ELSE                                                              235
       CHAT=8.D0*PI*PI*FAK(J+K+M+4)*(0.5D0/LAMBDA)**(J+K+M+5)           236
     &      *1.D0/(M+2.D0)*(1.D0/(K+1.D0)-1.D0/(K+M+3.D0))              237
      ENDIF                                                             238
      END                                                               239

      DOUBLE PRECISION FUNCTION WHAT(J,K,M)                             240
      IMPLICIT DOUBLE PRECISION (A-H,O-Z)                               241
      DOUBLE PRECISION NN                                               242
      WHAT=NN(J,K,M-1)                                                  243
      END                                                               244

      DOUBLE PRECISION FUNCTION FAK(N)                                  245
      IMPLICIT DOUBLE PRECISION (A-H,O-Z)                               246
      FAK=1.D0                                                          247
      DO 10 I=2,N                                                       248
       FAK=FAK*I                                                        249
   10 CONTINUE                                                          250
      END                                                               251
```

Fig. 16.5. FUNCTION subroutines **NN**, **THAT**, **CHAT**, **WHAT** and **FAK**

Figure 16.7 shows the numerical portion of the main program KAP16, the notation used being found in Fig. 16.6. In line 312 we determine $\hbar^2/(2m)$ from the formula

$$\frac{\hbar^2}{2m} = \frac{1}{2}\frac{\hbar^2}{me^2}e^2 = \frac{1}{2}r_0 e^2 \quad , \tag{16.48}$$

with r_0 from (16.6) and $e^2 = 14.40\,\text{eV\AA}$. In line 313 we obtain λ from (16.20).

The two nested DO-loops, beginning at lines 314 and 318, are used to compute the matrix elements $N_{nn'}$, $\tilde{T}_{nn'}$, $\tilde{C}_{nn'}$, $\tilde{W}_{nn'}$ and $\tilde{H}_{nn'}$. First the indices j, k, m and j', k', m' are stored away (lines 315 to 317 and 319 to 321) and the numbers J, K and M are computed from (16.22) (lines 322 to 324). Then from line 325 the matrix elements are determined from the following formulae: $N_{nn'}$ from (16.28) in line 325, $\tilde{T}_{nn'}$ from (16.35) in line 327, $\tilde{C}_{nn'}$ from (16.31) in line 329, $\tilde{W}_{nn'}$ from (16.33) in line 331 and $\tilde{H}_{nn'}$ from (16.37) in line 333 (plus continuation line 334). By taking the index NP only from 1 to N, instead of from 1 to NDIM (see line 318 and also line 314), we ensure that only the lower triangular matrix is computed. Then by interchanging the row and column indices we obtain the full symmetric matrices (lines 326, 328, 330, 332 and 335). The variable IEE in line 331 includes (IEE=1) or excludes (IEE=0) the electron-electron interaction.

The eigenvalue equation (16.19) is then solved by calling EIGENX in line 338. After transfer of the matrices $(\tilde{H}_{nn'})$ and $(N_{nn'})$ EIGENX produces the energy eigenvalues E_j in the array E and the eigenvectors $(\tilde{a}_n)_j$ in the form

Physical title	FORTRAN title	Physical title	FORTRAN title
$\hbar^2/(2m)$	H2M	N_{JKM}	NN(J,K,M)
e^2	EQUAD	$\hat{C}_{JKM}$	CHAT(J,K,M)
r_0	RO	$\hat{T}_{jkmj'k'm'}$	THAT(J1,K1,M1,
λ	LAMBDA		J2,K2,M2)
Z	Z	$\hat{W}_{JKM}$	WHAT(J,K,M)
α	ALPHA	$N_{nn'}$	NMAT(N,NP)
D	NDIM	$\tilde{T}_{nn'}$	TMAT(N,NP)
j	J1,JN(N)	$\tilde{C}_{nn'}$	CMAT(N,NP)
k	K1,KN(N)	$\tilde{W}_{nn'}$	WMAT(N,NP)
m	M1,MN(N)	$\tilde{H}_{nn'}$	HMAT(N,NP)
j'	J2,JN(NP)	E_j	E(J)
k'	K2,KN(NP)	$\tilde{a}_{nj}$	AMAT(N,J)
m'	M2,MN(NP)	$\langle T \rangle$	ERWT
J	J	$\langle C \rangle$	ERWC
K	K	$\langle W \rangle$	ERWW
M	M	$\langle H \rangle$	ERWH

Fig. 16.6. Notation used in main program KAP16

```
      PROGRAM KAP16                                                    300

      IMPLICIT DOUBLE PRECISION (A-H,O-Z)                              301
      DOUBLE PRECISION NN, NMAT, NORM                                  302
      INTEGER Z                                                        303
      PARAMETER (NMAX=20, EQUAD=14.40D0, R0=0.52917D0)                 304
      DIMENSION JN(NMAX), KN(NMAX), MN(NMAX)                           305
      DIMENSION E(NMAX), HF4(NMAX), HF5(NMAX), HF6(NMAX)               306
      DIMENSION HF1(NMAX,NMAX), HF2(NMAX,NMAX), HF3(NMAX,NMAX)         307
      DIMENSION CMAT(NMAX,NMAX), TMAT(NMAX,NMAX), WMAT(NMAX,NMAX)      308
      DIMENSION AMAT(NMAX,NMAX), HMAT(NMAX,NMAX), NMAT(NMAX,NMAX)      309
      DOUBLE PRECISION LAMBDA                                          310
      COMMON /LAMCOM/LAMBDA                                            311
*
      (Input: Z,IEE,NDIM,JN,KN,MN,ALPHA)

      H2M=0.5D0*R0*EQUAD                                               312
      LAMBDA=Z/(R0*ALPHA)                                              313
      DO 30 N=1,NDIM                                                   314
        J1=JN(N)                                                       315
        K1=KN(N)                                                       316
        M1=MN(N)                                                       317
        DO 20 NP=1,N                                                   318
          J2=JN(NP)                                                    319
          K2=KN(NP)                                                    320
          M2=MN(NP)                                                    321
          J=J1+J2                                                      322
          K=K1+K2                                                      323
          M=M1+M2                                                      324
          NMAT(N,NP)=NN(J,K,M)                                         325
          NMAT(NP,N)=NMAT(N,NP)                                        326
          TMAT(N,NP)=H2M*THAT(J1,K1,M1,J2,K2,M2)                       327
          TMAT(NP,N)=TMAT(N,NP)                                        328
          CMAT(N,NP)=-Z*EQUAD*CHAT(J,K,M)                              329
          CMAT(NP,N)=CMAT(N,NP)                                        330
          WMAT(N,NP)=IEE*EQUAD*WHAT(J,K,M)                             331
          WMAT(NP,N)=WMAT(N,NP)                                        332
          HMAT(N,NP)=CMAT(N,NP)+WMAT(N,NP)                             333
     &                         +TMAT(N,NP)                             334
          HMAT(NP,N)=HMAT(N,NP)                                        335
20      CONTINUE                                                       336
30    CONTINUE                                                         337
      CALL EIGENX(HMAT,NMAT,NMAX,NDIM,E,AMAT,HF1,HF2,HF3,HF4,HF5,HF6)  338
      ERWT=0.D0                                                        339
      ERWC=0.D0                                                        340
      ERWW=0.D0                                                        341
      DO 50 N=1,NDIM                                                   342
        DO 40 NP=1,NDIM                                                343
          ERWT=ERWT+AMAT(N,1)*TMAT(N,NP)*AMAT(NP,1)                    344
          ERWC=ERWC+AMAT(N,1)*CMAT(N,NP)*AMAT(NP,1)                    345
          ERWW=ERWW+AMAT(N,1)*WMAT(N,NP)*AMAT(NP,1)                    346
40      CONTINUE                                                       347
50    CONTINUE                                                         348
      ERWH=ERWT+ERWC+ERWW                                              349
*
      (Output: E,ERWT,ERWC,ERWW,ERWH,AMAT)

      END                                                              350
```

Fig. 16.7. Numerical portion of main program KAP16

of a matrix of coefficients stored in the two-dimensional array `AMAT`. The first index of `AMAT` indicates the individual components of a vector, whereas the second index refers to the corresponding eigenvalue. The first column of `AMAT` contains the eigenvector $(\tilde{a}_n)$ of the ground state ψ. The eigenvectors calculated by the subroutine `EIGEN` are normalised to unity. Accordingly the eigenvectors calculated from (16.45) by the subroutine `EIGENX` are normalised so that

$$\langle\psi|\psi\rangle = \sum_{n,n'} \tilde{a}_n N_{nn'} \tilde{a}_{n'} = 1 \quad . \tag{16.49}$$

In the lines 344 to 346 the expectation values of T, C and W are calculated,

$$\langle\psi|T|\psi\rangle = \sum_{n,n'} \tilde{a}_n \tilde{T}_{nn'} \tilde{a}_{n'} \,, \quad \langle\psi|C|\psi\rangle = \sum_{n,n'} \tilde{a}_n \tilde{C}_{nn'} \tilde{a}_{n'} \,,$$

$$\langle\psi|W|\psi\rangle = \sum_{n,n'} \tilde{a}_n \tilde{W}_{nn'} \tilde{a}_{n'} \quad . \tag{16.50}$$

Finally we form the sum

$$\langle H \rangle = \langle T \rangle + \langle C \rangle + \langle W \rangle \quad . \tag{16.51}$$

Since the expectation value of H in the ground state must be identical with the lowest eigenvalue we thereby get a little check on the calculation.

The following quantities have to be input: the nuclear charge number Z, the dimension D of the basis, the indices j, k, m of the basis states and the nonlinear variation parameter α. The output is limited to the data for the ground state. One obtains the energy eigenvalue E and the expectation values of the kinetic energy T, the electron-nucleus interaction C, the electron-electron interaction W and the Hamilton operator H, together with the components of the eigenvector $(\tilde{a}_n)$. In Fig. 16.8 we see as an example the screen display for Hylleraas's original calculation.

```
Results for the ground state, i.e. for the lowest eigenstate
(recall: ALPHA = 1.101):

Energy eigenvalue   E = -78.982216 eV

Expectation values:
   Kinetic energy                        <T> =    79.009224 eV
   Electron-nucleus interaction          <C> =  -183.769615 eV
   Electron-electron interaction         <W> =    25.778174 eV
   Hamilton operator                     <H> =   -78.982216 eV

Basis states used and corresponding components of the eigenvector:

              |   j   k   m     Components of the eigenvector
   ---------------------------------------------------------
   1. State   |   0   0   0            8.984912
   2. State   |   0   2   0            4.196493
   3. State   |   0   0   1            4.963162
```

Fig. 16.8. Screen display for $\alpha = 1.101$ with the three basis states used by Hylleraas

The program here summarised is to be found on the diskette in the file
KAP-16.FOR.

16.4 Exercises

16.4.1 The program enables one to treat the electron-electron interaction in first order perturbation theory. For this one sets $\alpha = 1$ and uses only the basis state with the indices $(j, k, m) = (0, 0, 0)$. How much is the binding energy influenced by the electron-electron interaction in this approximation?

16.4.2 The program enables one to carry out a simple Ritz variational calculation. One uses only the basis state with the indices $(j, k, m) = (0, 0, 0)$ and varies the parameter α. At which value of α does the minimum lie? By how much is the binding energy decreased, in comparison with the perturbation calculation of the first order?

16.4.3 Use all basis states with $j + k + m \leq 3$. By how much is the binding energy decreased in comparison with Hylleraas's calculation with the three basis states (0,0,0), (0,0,1) and (0,2,0); and by how much in comparison with the perturbation calculation of the first order?

16.5 Solutions to the Exercises

16.5.1 The binding energy is reduced by the expectation value $\langle W \rangle = 34.016\,\text{eV}$ of the electron-electron interaction potential. The first order perturbation calculation produces for the ground state energy of helium the value $-74.834\,\text{eV}$.

16.5.2 The minimum lies at $\alpha = 1.185$. For the ground state energy of helium one obtains the value $-77.492\,\text{eV}$. Compared with the first order perturbation calculation one gains $2.658\,\text{eV}$.

16.5.3 With the three basis states used by Hylleraas the minimum for the ground state energy lies at $\alpha = 1.101$. The ground state energy has the value $-78.982216\,\text{eV}$. With $j + k + m \leq 3$ one obtains 13 basis states. The minimum of the ground state energy lies at $\alpha = 1.050$ and has the value $-79.015104\,\text{eV}$. Compared with the historic calculation of Hylleraas the decrease is only $0.0329\,\text{eV}$. Compared with the first order perturbation calculation the decrease is about $4.2\,\text{eV}$.

In the Hamilton operator (16.29) we have neglected the motion of the nucleus, the spin-orbit interaction and relativistic effects. Even if we could solve the Schrödinger equation exactly, our ground state energy would therefore differ from the experimental one by a few hundredths of an eV. Since the experimental value for the ground state energy is $79.0\,\text{eV}$, we have therefore already achieved the best possible agreement.

17. The Spherical Harmonics

17.1 Formulation of the Problem

In Chaps. 13, 14 and 15 we have studied the motion of a particle of mass m under the influence of a spherically symmetric potential $V(r)$. With such a potential the forces act only in the direction of the line from the origin of coordinates to the position of the particle, i.e. in a system of polar coordinates (r, ϑ, φ) the forces act only in the direction of the coordinate r. The motion of the particle in its angular degrees of freedom ϑ and φ is a force-free motion. As we shall see, in this case a differential equation containing only these angular coordinates, and hence not coupled with an equation in r, can be separated from the Schrödinger equation. The solutions of this differential equation in ϑ and φ can be interpreted as wave functions on the unit sphere. One obtains a discrete set of eigensolutions if one imposes the condition that the wave functions close on each circuit of the unit sphere, i.e. that the wave functions are single-valued functions of ϑ and φ. As a normalisation condition one specifies that the absolute square of the wave function integrated over the spherical surface shall be equal to unity. These normalised eigensolutions are called the spherical harmonics and denoted by $Y_{l,m}(\vartheta, \varphi)$. We shall now consider these functions in more detail.

In quantum mechanics various paths lead to the differential equation for the spherical harmonics. One path leads via the separation of the Schrödinger equation. As already mentioned in Chap. 13, the Schrödinger equation for the motion of a point mass under the influence of a spherically symmetric potential is

$$\left(-\frac{\hbar^2}{2m}\Delta + V(r)\right)\psi(\mathbf{r}) = E\psi(\mathbf{r}) \quad . \tag{17.1}$$

Introducing polar coordinates (r, ϑ, φ) we tentatively express the wave function $\psi(\mathbf{r})$ in the form:

$$\psi(r, \vartheta, \varphi) = R(r)Y(\vartheta, \varphi) \quad . \tag{17.2}$$

The Laplace operator Δ in polar coordinates is

$$\Delta = \frac{1}{r^2}\frac{\partial}{\partial r}\left(r^2\frac{\partial}{\partial r}\right) + \frac{1}{r^2 \sin\vartheta}\frac{\partial}{\partial \vartheta}\left(\sin\vartheta\frac{\partial}{\partial \vartheta}\right) + \frac{1}{r^2 \sin^2\vartheta}\frac{\partial^2}{\partial \varphi^2} \quad . \tag{17.3}$$

Substituting (17.2) and (17.3) in (17.1), one sees that the equation separates.

There are terms which depend only on the coordinate r, and terms which depend only on ϑ and φ. When the latter are brought to the right-hand side we obtain

$$\frac{1}{R}\frac{\partial}{\partial r}\left(r^2\frac{\partial R}{\partial r}\right) + \frac{2m}{\hbar^2}r^2(E-V(r))$$
$$= -\frac{1}{Y}\left(\frac{1}{\sin\vartheta}\frac{\partial}{\partial\vartheta}\left(\sin\vartheta\frac{\partial Y}{\partial\vartheta}\right) + \frac{1}{\sin^2\vartheta}\frac{\partial^2 Y}{\partial\varphi^2}\right) . \tag{17.4}$$

Since the two sides depend on different coordinates, each side must be equal to a constant, which we shall call λ. We accordingly obtain the two equations

$$\frac{1}{R}\frac{\partial}{\partial r}\left(r^2\frac{\partial R}{\partial r}\right) + \frac{2m}{\hbar^2}r^2(E-V(r)) = \lambda \tag{17.5}$$

and

$$-\frac{1}{Y}\left(\frac{1}{\sin\vartheta}\frac{\partial}{\partial\vartheta}\left(\sin\vartheta\frac{\partial Y}{\partial\vartheta}\right) + \frac{1}{\sin^2\vartheta}\frac{\partial^2 Y}{\partial\varphi^2}\right) = \lambda . \tag{17.6}$$

Equation (17.5) simplifies if we define the wave function $u(r)$ by

$$R(r) = \frac{1}{r}u(r) . \tag{17.7}$$

Substitution of (17.7) and (17.8) into (17.5) leads, after a little manipluation, to the radial Schrödinger equation, which we have already considered in Chap. 13.

The solutions of (17.6) are well known in mathematics. The condition that the functions close on each circuit of the unit sphere can be satisfied only if

$$\lambda = l(l+1) \quad \text{with} \quad l = 0, 1, 2, 3, \ldots . \tag{17.8}$$

After multiplication by $\sin^2\vartheta$ the differential operator occurring in (17.6) separates into one term which depends only on ϑ, and one term which depends only on φ. With the tentative formulation

$$Y(\vartheta,\varphi) = \Theta(\vartheta)\Phi(\varphi) \tag{17.9}$$

(17.6) separates into the two equations

$$-\frac{1}{\Phi}\frac{d^2\Phi}{d\varphi^2} = \nu \quad \text{and} \tag{17.10}$$

$$\frac{1}{\Theta}\sin\vartheta\frac{d}{d\vartheta}\left(\sin\vartheta\frac{d\Theta}{d\vartheta}\right) + \lambda\sin^2\vartheta = \nu . \tag{17.11}$$

From (17.10) one deduces that the constant ν must be equal to the square of an integer m,

$$\nu = m^2 \quad \text{with} \quad m = 0, \pm 1, \pm 2, \pm 3, \ldots , \tag{17.12}$$

because only then will $\Phi(\varphi)$ be a single-valued function of the angle φ. The (normalised) solutions of (17.10) are

$$\Phi_m(\varphi) = \frac{1}{(2\pi)^{1/2}} e^{im\phi} \quad . \tag{17.13}$$

Before one tries to solve (17.11) it is helpful to modify the equation. One multiplies (17.11) by $\Theta/\sin^2\vartheta$, transforms from the variable ϑ to the variable $w = \cos\vartheta$ and then seeks solutions $P(w)$ with the notation

$$P(w) = P(\cos\vartheta) = \Theta(\vartheta) \quad . \tag{17.14}$$

The differential equation for $P(w)$ is

$$\frac{d}{dw}\left((1-w^2)\frac{dP_l^m(w)}{dw}\right) + \left(l(l+1) - \frac{m^2}{1-w^2}\right)P_l^m(w) = 0 \quad . \tag{17.15}$$

Here we have already applied the conditions (17.8) and (17.12), under which the solutions of the differential equation are single-valued, and have labelled the solutions $P_l^m(w)$ accordingly. The functions $P_l^m(w)$ are called the associated Legendre functions. The spherical harmonics accordingly become

$$Y_{l,m}(\vartheta,\varphi) = N_{lm} P_l^m(\cos\vartheta)\Phi_m(\varphi) \quad , \tag{17.16}$$

where N_{lm} is a normalising constant.

The relation between the quantum numbers l, m and physical quantities becomes apparent if we consider the second path for obtaining the differential equation of the spherical harmonics.

We consider the operator for the angular momentum

$$\mathbf{L} = (L_x, L_y, L_z) \tag{17.17}$$

with

$$L_x = -i\hbar\left(y\frac{\partial}{\partial z} - z\frac{\partial}{\partial y}\right) \quad , \quad L_y = -i\hbar\left(z\frac{\partial}{\partial x} - x\frac{\partial}{\partial z}\right) \quad ,$$

$$L_z = -i\hbar\left(x\frac{\partial}{\partial y} - y\frac{\partial}{\partial x}\right) \quad . \tag{17.18}$$

Transformation to polar coordinates (r,ϑ,φ) gives

$$L_z = -i\hbar\frac{\partial}{\partial\varphi} \quad \text{and} \tag{17.19}$$

$$L^2 = -\hbar^2\left(\frac{1}{\sin\vartheta}\frac{\partial}{\partial\vartheta}\left(\sin\vartheta\frac{\partial}{\partial\vartheta}\right) + \frac{1}{\sin^2\vartheta}\frac{\partial^2}{\partial\varphi^2}\right) \quad . \tag{17.20}$$

The eigenvalue equation set up with the operator (17.20) is identical with equation (17.6). It follows that the solutions of (17.6) are eigenstates of the square of the angular momentum operator with the eigenvalues $\hbar^2\lambda = \hbar^2 l(l+1)$. One therefore calls l the angular momentum quantum number. The eigenvalue equation set up by the operator (17.19) has the functions (17.13) as solutions. Their eigenvalues are $\hbar m$. One therefore calls m the quantum number of the z-component of the angular momentum, or the magnetic quantum number.

17.2 Numerical Method

One can derive the associated Legendre functions $P_l^m(w)$ for $l \geq m \geq 0$ and $|w| \leq 1$ from the Legendre polynomials $P_l(w)$. We have[1] [17.1, 2]

$$P_l^m(w) = (1-w^2)^{m/2} \frac{d^m}{dw^m} P_l(w) \quad . \tag{17.21}$$

The Legendre polynomial $P_l(w)$ can be obtained from the Rodrigues formula

$$P_l(w) = \frac{1}{2^l l!} \frac{d^l}{dw^l} (w^2-1)^l \quad . \tag{17.22}$$

Since m enters quadratically into equation (17.15), one obtains for $m < 0$ the solutions $P_l^{|m|}(w)$. Thus in our definition (17.21) we can assume m to be positive. From (17.21) it also follows that

$$P_l^m(w) = 0 \quad \text{for} \quad 0 \leq l < m \quad , \tag{17.23}$$

since in this case a polynomial of degree $2 \times l$ is differentiated more than $(2 \times l)$ times.

The case $l = m$ is especially simple, since then the differentiation leaves only the term

$$P_m^m(w) = (1-w^2)^{m/2} \frac{d^m}{dw^m} P_m(w) = \frac{(2m)!}{2^m m!} \overline{w}^m \tag{17.24}$$

where we have set

$$\overline{w} = (1-w^2)^{1/2} \quad . \tag{17.25}$$

The associated Legendre functions can be calculated by a recursion formula [17.1, 2],

$$P_{l+1}^m(w) = \frac{1}{l-m+1}((2l+1)w P_l^m(w) - (l+m)P_{l-1}^m(w)) \quad . \tag{17.26}$$

Starting from P_m^m, one obtains the polynomials P_{m+1}^m, P_{m+2}^m, ... up to the polynomial with the required value of l. As starting functions for the recursion we have P_m^m from (17.24) and P_{m+1}^m from the formula

$$P_{m+1}^m(w) = (2m+1)w P_m^m(w) \quad , \tag{17.27}$$

which follows from (17.26) and (17.23). Apart from its simplicity, (17.26) has the advantage of good numerical stability.

The factor N_{lm} occurring in (17.16) is, apart from a phase factor of modulus unity, determined by the normalisation condition

$$\int_0^\pi d\vartheta \int_0^{2\pi} |Y_{l,m}(\vartheta,\varphi)|^2 \sin\vartheta \, d\varphi = 1 \quad . \tag{17.28}$$

---

[1] In [17.2] the definition contains an additional phase factor $(-1)^m$.

With the usual choice of phase factor the spherical harmonics [17.1] are

$$Y_{l,m}(\vartheta,\varphi) = c_m \left(\frac{(2l+1)(l-|m|)!}{4\pi(l+|m|)!}\right)^{1/2} P_l^{|m|}(\cos\vartheta)e^{im\varphi} , \qquad (17.29)$$

where $c_m = (-1)^m$ for $m \geq 0$ and $c_m = 1$ for $m < 0$. Unfortunately the phase factor c_m is not chosen the same in all text-books.

The first four spherical harmonics are

$$Y_{0,0} = \frac{1}{(4\pi)^{1/2}} , \quad Y_{1,1} = -\left(\frac{3}{8\pi}\right)^{1/2} \sin\vartheta e^{i\varphi} ,$$

$$Y_{1,0} = \left(\frac{3}{4\pi}\right)^{1/2} \cos\vartheta , \quad Y_{1,-1} = \left(\frac{3}{8\pi}\right)^{1/2} \sin\vartheta e^{-i\varphi} . \qquad (17.30)$$

17.3 Programming

We compute the associated Legendre functions in the FUNCTION subprogram PLM (Figs. 17.1 and 17.2). One has to transfer to this the angle ϑ in radians, the angular momentum quantum number l and the magnetic quantum number m. One obtains the function values $P_l^m(\cos\vartheta)$. For $m < 0$ the program stops.

After the calculation of $w = \cos\vartheta$ and $\overline{w} = \sin\vartheta$ (lines 104 and 108), P_m^m is calculated from (17.24) (line 109). For the calculation of the factorial we use the FUNCTION subprogram FAK from Chap. 16. The case $m = 0$ is treated as a special case (line 105), because when $\overline{w} = 0$ the term $\overline{w}^m = 0^0$ is not defined. We have $P_0^0(w) = 1$ (line 106). In line 112 (17.27) is used, and in the DO-loop from line 114 all the remaining polynomials are computed from the recursion (17.26).

Figure 17.4 shows the main program KAP17. The notation employed is to be found in Fig. 17.3. For the angular range of ϑ we take the interval $[0,\pi]$, which we divide into i_π sub-intervals of size

$$h = \frac{\pi}{i_\pi} . \qquad (17.31)$$

The mesh points are

$$\vartheta_i = ih , \quad 0 \leq i \leq i_\pi . \qquad (17.32)$$

Mathematical title	FORTRAN title	Mathematical title	FORTRAN title
l	L	w	W
m	M	$\overline{w}$	WBAR
$m!$	FAK(M)	ϑ	THETA
		$P_l^m(w)$	PLM

Fig. 17.1. Notation used in the FUNCTION subprogram PLM

```
        DOUBLE PRECISION FUNCTION PLM(THETA,L,M)                100
        IMPLICIT DOUBLE PRECISION (A-H,O-Z)                     101
        IF (M.LT.0)                                             102
       & STOP 'Error in FUNCTION PLM: 0 ≤ M ≤ L violated'       103
        W=COS(THETA)                                            104
        IF(M.EQ.0) THEN                                         105
          P0=1.D0                                               106
        ELSE                                                    107
          WBAR=SIN(THETA)                                       108
          P0=FAK(2*M)/(2.D0**M*FAK(M))*WBAR**M                  109
        ENDIF                                                   110
        IF (M.LE.L-1) THEN                                      111
          P1=(2*M+1)*W*P0                                       112
          IF (M.LE.L-2) THEN                                    113
            DO 10 I=M+1,L-1                                     114
              P2=1.D0/(I-M+1)*((2*I+1)*W*P1-(I+M)*P0)           115
              P0=P1                                             116
              P1=P2                                             117
10        CONTINUE                                              118
          END IF                                                119
          PLM=P1                                                120
        ELSE IF (M.EQ.L) THEN                                   121
          PLM=P0                                                122
        ELSE                                                    123
          PLM=0                                                 124
        ENDIF                                                   125
        END                                                     126
```

Fig. 17.2. FUNCTION subprogram PLM

In the DO-loop of lines 209 to 215 the mesh points are determined, and the real and imaginary parts of $Y_{lm}(\vartheta_i,\varphi)$ are computed from (17.29), where use is made of the relation

$$e^{im\varphi} = \cos(m\varphi) + i\sin(m\varphi) \quad . \tag{17.33}$$

The values for π and i_π are specified in a PARAMETER instruction (line 203). The angular momentum quantum number l, the magnetic qunatum number m and the azimuth angle φ are input. As output we obtain the real part $\mathrm{Re}\,[Y_{l,m}(\vartheta,\varphi)]$ and the imaginary part $\mathrm{Im}\,[Y_{l,m}(\vartheta,\varphi)]$ of the corresponding

Physical title	FORTRAN title	Physical title	FORTRAN title
π	PI	l	L
i_π	IPI	m	M
h	H	$|m|$	MABS
c_m	CM	$l!$	FAK(L)
φ	PHI	$P_l^m(\cos\vartheta_i)$	PLM(THETA(I),L,M)
ϑ_i	THETA(I)	$\mathrm{Re}[Y_{lm}(\vartheta_i,\varphi)]$	REY(I)
		$\mathrm{Im}[Y_{lm}(\vartheta_i,\varphi)]$	IMY(I)

Fig. 17.3. Notation employed in main program KAP17

```
      PROGRAM KAP17                                             200

      IMPLICIT DOUBLE PRECISION (A-H,O-Z)                       201
      DOUBLE PRECISION IMY                                      202
      PARAMETER (IPI=180, PI=3.1415926535D0)                    203
      DIMENSION THETA(0:IPI), REY(0:IPI), IMY(0:IPI)            204
*     (Input of L, M, PHI)

      H=PI/IPI                                                  205
      MABS=ABS(M)                                               206
      CM=1.D0                                                   207
      IF (M.GE.0) CM=(-1.D0)**M                                 208
      DO 10 I=0,IPI                                             209
      THETA(I)=I*H                                              210
      REY(I)=CM*SQRT((2.D0*L+1.D0)*FAK(L-MABS)/(4.D0*PI*FAK(L   211
     &       +MABS)))*PLM(THETA(I),L,MABS)*COS(M*PHI)           212
      IMY(I)=CM*SQRT((2.D0*L+1.D0)*FAK(L-MABS)/(4.D0*PI*FAK(L   213
     &       +MABS)))*PLM(THETA(I),L,MABS)*SIN(M*PHI)           214
10    CONTINUE                                                  215

*     (Output of REY, IMY)

      END                                                       216
```

Fig. 17.4. Numerical portion of main program KAP17

spherical harmonic in the region $0 \leq \vartheta \leq \pi$. The output is given in graphical form and optionally also in numerical form.

The program here summarised is to be found on the diskette in the file KAP-17.FOR.

17.4 Exercises

17.4.1 The spherical harmonics are the angular parts of the quantum mechanical wave functions for fixed eigenvalues of L^2 and L_z. The probability density calculated with such wave functions is a function of the polar angle ϑ, but does not depend on the azimuthal angle φ. The classical analogue to a quantum mechanical motion with specified values for L^2 and L_z is a circular orbit on the (unit) sphere. We can establish a connection between the classical and the quantum mechanical probability densities, by averaging over all the classical orbits which have equal inclinations to the equator. As a demonstration we can take the orbit of an artificial earth satellite. As the earth spins under the orbit, there is an average position probability. For all places on the same degree of latitude there is an equal probability of the satellite passing overhead.

Study the connection between the classical motion and the quantum mechanical wave functions in the following cases:

a) Classical orbit along the equator, spherical harmonics with $l = m$ ($\neq 0$).
b) Classical orbit over the poles, spherical harmonics with $l \neq 0$, $m = 0$.
c) Classical orbit between two circles of latitude, spherical harmonics with $l \neq 0$, $m \neq l$.

17.4.2 With the help of the numerical output study the behaviour of the spherical harmonics in the neighbourhood of the poles ($\vartheta \to 0°$), ($\vartheta \to 180°$). Can you formulate a simple rule?

17.5 Solutions to the Exercises

17.5.1 a) One obtains wave functions which, relative to the angle ϑ, are similar to the ones of zero-point oscillations. In fact the physical circumstances are also similar. Classically one would expect a delta function at the equator. Quantum mechanically a precise localisation is possible only with infinitely high quantum numbers. Figure 17.5 shows $Y_{ll}(\vartheta, 0)$ for $l = 2, 6$ and 10. It is noticeable that the localisation at the equator becomes sharper for larger angular momentum quantum number l. With regard to the angle φ all the spherical harmonics have the same simple behaviour $\exp(im\varphi)$.

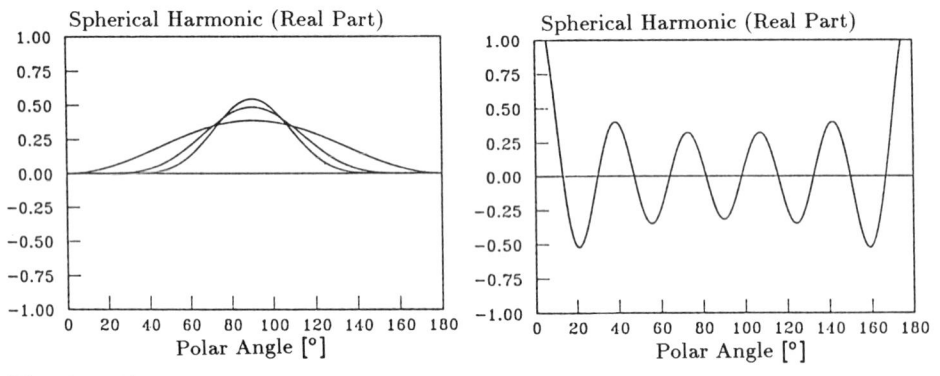

Fig. 17.5. The spherical harmonics $Y_{ll}(\vartheta, 0)$ for $l = 2, 6, 10$

Fig. 17.6. The spherical harmonic $Y_{10,0}(\vartheta, 0)$

b) With regard to the angle ϑ one obtains wave functions which have their greatest amplitude at the poles. With larger l-values one clearly sees a decrease of amplitude towards the equator (see Fig. 17.6). This behaviour corresponds to the fact that one encounters a satellite (in polar orbit) with greater probability in a locality in the polar neighbourhood than in a locality of equal area in the neighbourhood of the equator. The minima occurring in the absolute square of the spherical harmonics have no classical analogue.

c) The spherical harmonics have their greatest amplitudes in the neighbourhood of those circles of latitude which the classical orbit touches before turning back towards the equator, see Fig. 17.7. As with the oscillator functions in Chap. 14, there is here also a close connection between the absolute square of the quantum mechanical wave function and the probability density of a particle moving along the corresponding classical orbit.

17.5.2 In the neighbourhood of the pole the spherical harmonics $Y_{l,m}(\vartheta, \varphi)$ behave like $(\sin \vartheta)^m$. Since $\sin \vartheta$ denotes the distance from the z-axis, this behaviour is analogous to the behaviour $R_l(r) \to r^l$ of the radial solutions of the Schrödinger equation. In both cases, a centrifugal barrier determines the behaviour of the functions. Figure 17.7 shows how the spherical harmonic $Y_{8,4}(\vartheta, 0)$ vanishes at the poles.

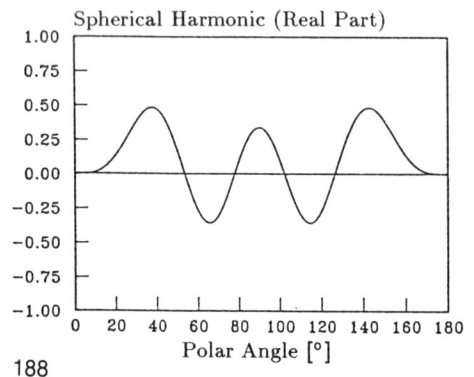

Fig. 17.7. The spherical harmonic $Y_{8,4}(\vartheta, 0)$

18. The Spherical Bessel Functions

18.1 Formulation of the Problem

We have already in several chapters studied the solutions of the one-particle Schrödinger equation

$$-\frac{\hbar^2}{2m}\Delta\psi(\mathbf{r}) + V(r)\psi(\mathbf{r}) = E\psi(\mathbf{r}) \ . \tag{18.1}$$

The equation is also valid, as we know, for the relative motion of two particles. In this case $\mathbf{r}$ is the separation distance vector and m the reduced mass of the two particles. For a spherically symmetric potential (18.1) can be solved relatively easily. One introduces spherical coordinates (r, ϑ, φ) and decomposes the wave function into angular momentum partial waves (see Chap. 17),

$$\psi(r,\vartheta,\varphi) = \sum_{l,m} c_{lm} \frac{1}{r} u_l(r) Y_{lm}(\vartheta,\varphi) \ . \tag{18.2}$$

The radial wave functions $u_l(r)$ can be calculated by the method given in Chap. 13, the spherical functions $Y_{lm}(\vartheta, \varphi)$ by the method given in Chap. 17.

We still lack something: in Chap. 13 we could calculate the radial wave function $u_l(r)$ only on a finite interval $[0, b]$. If we wish to determine the solution $\psi(r)$ in the whole configuration space, then we have to continue the radial wave functions into the interval $[b, \infty]$.

We assume that the potential $V(r)$ decreases strongly for large values of r and that we have chosen b already so large that $V(r)$ can be neglected for $r > b$. Then the radial Schrödinger equation (13.3) goes over in the interval $[b, \infty]$ into the free radial equation

$$-\frac{\hbar^2}{2m}\frac{d^2}{dr^2}\overline{u}_l(r) + \frac{\hbar^2}{2m}\frac{l(l+1)}{r^2}\overline{u}_l(r) = E\overline{u}_l(r) \ . \tag{18.3}$$

The solutions of this equation are well-known analytic functions. We can use them in order to continue into the interval $[b, \infty]$ the solutions $u_l(r)$ calculated in the interval $[0, b]$ by the Fox-Goodwin method.

The continuation of the solutions can be most simply demonstrated in the case $l = 0$. For $l = 0$ the general solution of (18.3) in the interval $[b, \infty]$ is

$$\overline{u}_0(r) = a_0 \sin(kr + \delta_0) = a_0[\cos(\delta_0)\sin(kr) + \sin(\delta_0)\cos(kr)] , \tag{18.4}$$

where

$$k = \frac{(2mE)^{1/2}}{\hbar} \qquad (18.5)$$

is the wave number.

At the interval limit $r = b$ the solution $u_0(r)$ of the radial equation (13.3) must go over with a continuous first derivative into the free solution $\bar{u}_0(r)$. However, the first derivative of the function $u_0(r)$ is not known to us. We have to calculate it from the function tabulated in the interval $[0, b]$. Instead of this, however, we can avoid the calculation of the derivative and simply equate the functions $u_0(r)$ and $\bar{u}_0(r)$ at the last two mesh points $b-h$ and b of the interval $[0, b]$:

$$\begin{aligned} u_0(b-h) &= a_0[\cos(\delta_0)\sin(k(b-h)) + \sin(\delta_0)\cos(k(b-h))] \\ u_0(b) &= a_0[\cos(\delta_0)\sin(kb) + \sin(\delta_0)\cos(kb)] \end{aligned} \qquad (18.6)$$

From these two equations a_0 and δ_0 can be calculated. First one obtains

$$\tan(\delta_0) = \frac{\sin(k(b-h))u_0(b) - \sin(kb)u_0(b-h)}{-\cos(k(b-h))u_0(b) + \cos(kb)u_0(b-h)} \qquad (18.7)$$

and hence finds δ_0 modulo π. One calls δ_0 the scattering phase shift of angular momentum $l = 0$. It has the following significance. The solution of the free radial equation (18.3) for $l = 0$, with physical boundary condition at the origin, is $a_0 \sin(kr)$, and δ_0 is the phase shift of the scattering solution compared with the free solution in the asymptotic region. If one has the phase shift δ_0 modulo π, then $\sin(\delta_0)$ and $\cos(\delta_0)$ are known apart from the sign, and one can, after specifying the sign, calculate the amplitude a_0 from equation (18.6). The amplitude a_0 depends on the normalisation of $u_0(r)$.

We have now to extend the treatment to the case $l \neq 0$. The general solution of equation (18.3) is

$$\bar{u}_l(r) = a_l kr[\cos(\delta_l)j_l(kr) - \sin(\delta_l)n_l(kr)] \qquad (18.8)$$

The functions $j_l(kr)$ and $n_l(kr)$ are called the regular and irregular spherical Bessel functions. The extraction of the factor kr in (18.8) is a convention and is part of the definition of the spherical Bessel functions. The spherical Bessel functions are radial functions of the type $R_l(r) = (1/r)\bar{u}_l(r)$, see (17.5) and (17.7). Regular means that the functions $j_l(kr)$ satisfy a physical boundary condition at the origin of coordinates, whereas the functions $n_l(kr)$ diverge at the origin and asymptotically have a phase shift of $-\frac{\pi}{2}$ relative to the regular functions.

The spherical Bessel functions are analytically well-known functions. We shall compute them and study them on the screen. For the computation we shall use recursion formulae, without deriving them. The interested reader will find the derivations in mathematical text-books under "Spherical Bessel Functions" or under "Solutions of Bessel's differential equation with half-integer index".

The most important formulae with accounts of their origin are also to be found in [18.1] and [18.2][1].

We can continue the radial solutions $u_l(r)$ into the region $r > b$ for $l \neq 0$ in the same way as for $l = 0$, if we replace equation (18.4) by (18.8). In order to determine δ_l and a_l we then only have to replace the functions $\sin(kr)$ and $\cos(kr)$ in (18.6) and (18.7) by $krj_l(kr)$ and $-krn_l(kr)$. We obtain the equations

$$u_l(b-h) = a_l k(b-h)[\cos(\delta_l)j_l(k(b-h)) - \sin(\delta_l)n_l(k(b-h))] ,$$
$$u_l(b) = a_l kb[\cos(\delta_l)j_l(kb) - \sin(\delta_l)n_l(kb)] \qquad (18.9)$$

and

$$\tan(\delta_l) = \frac{(b-h)j_l(k(b-h))u_l(b) - bj_l(kb)u_l(b-h)}{(b-h)n_l(k(b-h))u_l(b) - bn_l(kb)u_l(b-h)} . \qquad (18.10)$$

For $l = 0$ these equations become (18.6) and (18.7).

Also of interest to us is the asymptotic form of the spherical Bessel functions. For $r \to \infty$ we have

$$krj_l(kr) \to \sin(kr - l\tfrac{\pi}{2}) \qquad (18.11)$$

and

$$krn_l(kr) \to -\cos(kr - l\tfrac{\pi}{2}) . \qquad (18.12)$$

As already expressed in equation (18.8), both $krj_l(kr)$ and $krn_l(kr)$ satisfy the radial equation (18.3) of the free motion. In the asymptotic region, where the centrifugal potential vanishes, the functions $krj_l(kr)$ and $krn_l(kr)$ must be of the form (18.4). One realises that the phase shift $-l\tfrac{\pi}{2}$ originates from the centrifugal potential.

If we choose the matching value b in (18.9) very large, then we can substitute for the spherical Bessel functions their asymptotic forms (18.11) and (18.12). Applying (18.4) we then see that our asymptotic solution has the form

$$\overline{u}_l(r) \to a_l \sin(kr - l\tfrac{\pi}{2} + \delta_l) . \qquad (18.13)$$

This clarifies the meaning of the quantity δ_l for $l \neq 0$. It is the asymptotic phase difference between the physical solution of the Schrödinger radial equation with potential $V(r)$ and centrifugal potential, and the solution of the Schrödinger radial equation with centrifugal potential only.

18.2 Mathematical Method

As for the spherical harmonics in the previous chapter, we shall carry out the calculation of the Bessel functions by a recursion formula. Since the spherical Bessel functions depend only on the product kr, we shall write:

---

[1] In [18.2] $n_l(kr)$ is defined with the opposite sign.

$$x = kr \quad . \tag{18.14}$$

By comparison of (18.8) with (18.4) one sees that for $l = 0$ the following relations hold:

$$j_0(x) = \frac{\sin x}{x} \quad , \quad n_0(x) = -\frac{\cos x}{x} \quad . \tag{18.15}$$

For $l = 1$ we have [18.1]

$$j_1(x) = \frac{\sin x}{x^2} - \frac{\cos x}{x} \quad , \quad n_1(x) = -\frac{\cos x}{x^2} - \frac{\sin x}{x} \quad . \tag{18.16}$$

The subsequent spherical Bessel functions can be calculated from the recursion formula [18.1]

$$f_{l+1}(x) = \frac{2l+1}{x} f_l(x) - f_{l-1}(x) \tag{18.17}$$

where f_l denotes either j_l or n_l. The calculation of the regular spherical Bessel functions with this formula is not numerically stable near the origin for large values of l. At each step of the recursion it involves the subtraction of almost equal quantities, which accumulates and amplifies rounding errors. We shall therefore omit the interval $0 \le x \le 0.2$ from the graphical presentation. The numerical output will show the instability. The true profile of the function in the neighbourhood of the origin is quite simple. As one can see substituting a power series expansion into the recursion formula, one has for $x \to 0$

$$j_l(x) \to \frac{x^l}{1 \cdot 3 \cdot 5 \cdots (2l+1)} \quad , \tag{18.18a}$$

$$n_l(x) \to \frac{1 \cdot 3 \cdot 5 \cdots (2l-1)}{x^{l+1}} \quad . \tag{18.18b}$$

In the mathematical literature, instead of the spherical Bessel functions j_l one usually finds the Bessel functions J_ν, which are connected to the j_l by the relation

$$j_l(x) = \left(\frac{\pi}{2x}\right)^{1/2} J_{l+1/2}(x) \quad . \tag{18.19}$$

As one can show by substitution in (18.3) and subsequent manipulation, the J_ν satisfy the differential equation

$$x^2 J_\nu''(x) + x J_\nu'(x) + (x^2 - \nu^2) J_\nu(x) = 0 \quad , \tag{18.20}$$

which is known as Bessel's differential equation.

18.3 Programming

For the calculation of the regular and the irregular spherical Bessel functions we set up two FUNCTION subprograms. The first subprogram we call BESJ,

Physical title	FORTRAN title	Physical title	FORTRAN title
x	X	l $j_l(x)$	L BESJ

Fig. 18.1. Notation employed in FUNCTION subprogram BESJ

```
      DOUBLE PRECISION FUNCTION BESJ(X,L)                100
      IMPLICIT DOUBLE PRECISION (A-H,O-Z)                101
      IF(X.LE.0.D0) STOP                                 102
     &'Error in SUBROUTINE BESJ: x > 0 not satisfied'    103
      B0=SIN(X)/X                                        104
      BESJ=B0                                            105
      IF(L.GE.1) THEN                                    106
        B1=SIN(X)/(X*X)-COS(X)/X                         107
        BESJ=B1                                          108
        IF(L.GE.2) THEN                                  109
          DO 10 I=1,L-1                                  110
            BESJ=(2*I+1)/X*B1-B0                         111
            B0=B1                                        112
            B1=BESJ                                      113
 10       CONTINUE                                       114
        ENDIF                                            115
      ENDIF                                              116
      END                                                117
```

Fig. 18.2. FUNCTION subprogram BESJ

the second one BESN (Figs. 18.1 to 18.4). Since both programs have the same formal construction, and are moreover rather straightforward, we can discuss them together.

The spherical Bessel functions for $l = 0$ and $l = 1$ are calculated from (18.15) and (18.16) (lines 104 and 107, 204 and 207). For higher values of l the recursion formula (18.17) is applied (line 111, 211).

In the main program KAP18 (Figs. 18.5 and 18.6) we divide the interval $[0, b]$ as usual into a number i_b of sub-intervals, resulting in the mesh width

$$h = \frac{b}{i_b} \qquad (18.21)$$

(line 305) and determine the mesh points

$$x_i = ih \quad , \quad 1 \le i \le i_b \qquad (18.22)$$

Physical title	FORTRAN title	Physical title	FORTRAN title
x	X	l $n_l(x)$	L BESN

Fig. 18.3. Notation employed in FUNCTION subprogram BESN

193

```
      DOUBLE PRECISION FUNCTION BESN(X,L)                200

      IMPLICIT DOUBLE PRECISION (A-H,O-Z)                201
      IF(X.LE.0.D0) STOP                                 202
     &'Error in SUBROUTINE BESN: x > 0 not satisfied'    203
      B0=-COS(X)/X                                       204
      BESN=B0                                            205
      IF(L.GE.1) THEN                                    206
        B1=-COS(X)/(X*X)-SIN(X)/X                        207
        BESN=B1                                          208
        IF(L.GE.2) THEN                                  209
          DO 10 I=1,L-1                                  210
            BESN=(2*I+1)/X*B1-B0                         211
            B0=B1                                        212
            B1=BESN                                      213
10        CONTINUE                                       214
        ENDIF                                            215
      ENDIF                                              216
      END                                                217
```

Fig. 18.4. FUNCTION subprogram BESN

Physical title	FORTRAN title	Physical title	FORTRAN title
i_b	IB	l	L
b	B	$j_l(x_i)$	BESJ(L)
h	H	$n_l(x_i)$	BESN(L)
x_i	X(I)		

Fig. 18.5. Notation used in main program KAP18

```
      PROGRAM KAP18                                      300

      IMPLICIT DOUBLE PRECISION (A-H,O-Z)                301
      DOUBLE PRECISION JL, NL                            302
      PARAMETER (IB=200, B=20.D0)                        303
      DIMENSION JL(IB), NL(IB), X(IB)                    304
*     (Input: L)

      H=B/IB                                             305
      DO 20 I=1,IB                                       306
        X(I)=I*H                                         307
        H1=X(I)                                          308
        JL(I)=BESJ(H1,L)                                 309
        NL(I)=BESN(H1,L)                                 310
20    CONTINUE                                           311
*     (Output: JL, NL)

      END                                                312
```

Fig. 18.6. Numerical portion of main program KAP18

(line 307). The calculation of the function values of the spherical Bessel functions at the mesh points then follows by calling the FUNCTION subprograms BESJ and BESN (lines 309 and 310).

The values $b = 20$ and $i_b = 200$ are specified in a PARAMETER instruction (line 303). The only input is the angular momentum quantum number l. As output one obtains the spherical Bessel functions $j_l(x)$ and $n_l(x)$ pertaining to this value of l.

The program here summarised is to be found on the diskette in the file KAP-18.FOR.

18.4 Exercises

18.4.1 Display and study the regular and irregular spherical Bessel functions for $0 \leq l \leq 10$ on the screen.

18.4.2 Test the accuracy of the numerically displayed function values for $j_l(x)$ and $n_l(x)$ in the region $0 \leq x \leq 1$ by comparison with (18.18).

18.5 Solutions to the Exercises

18.5.1 Figure 18.7 shows the regular and irregular spherical Bessel functions for $l = 0, 1, 2$ and 10.

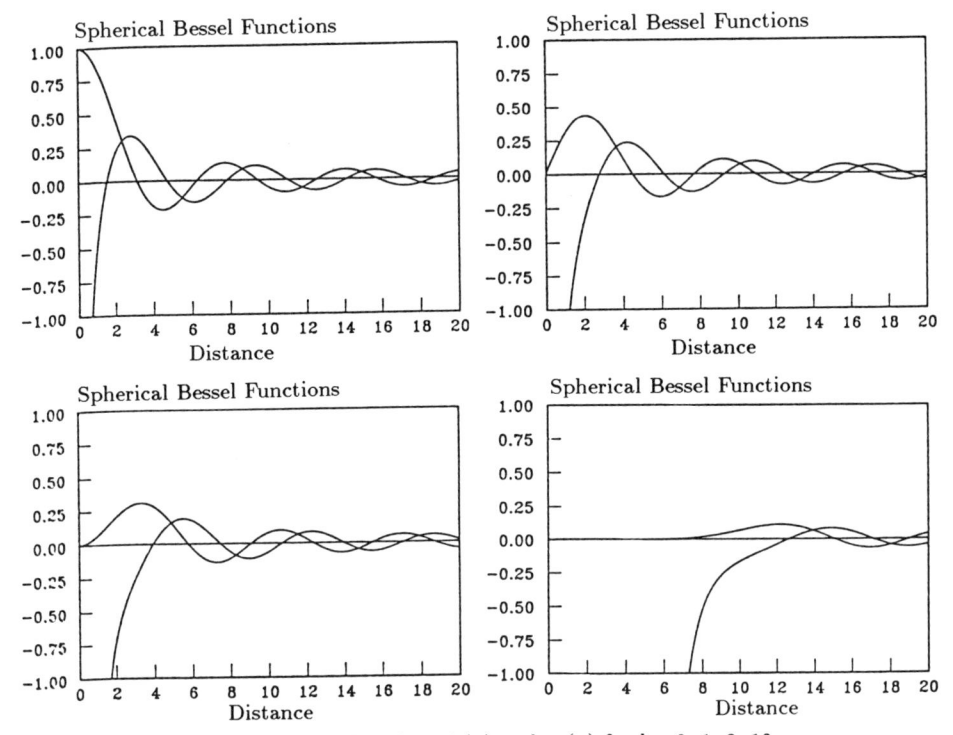

Fig. 18.7. The spherical Bessel functions $j_l(x)$ and $n_l(x)$ for $l = 0, 1, 2, 10$

18.5.2 Comparison of the numerical function values with the values deduced from (18.18) shows no detectable error in the irregular functions for $l \gg 0$ and $x \ll 1$; deviations of 0.1 % at $x = 0.2$, 0.5 % at $x = 0.4$ and 0.9 % at $x = 0.6$ for $l = 10$ only reveal that (18.18) is increasingly inaccurate for larger values of x. The regular functions, on the other hand, become so inaccurate, e.g. at $x = 0.2$ and $l = 10$, that it would even perturb the graphical output. The graphical output therefore begins only at $x > 0.2$. For larger values of x the results become very accurate both for the regular and the irregular spherical Bessel functions. One can check this, for example, by the relation [18.1]

$$n_{l-1}(x)j_l(x) - n_l(x)j_{l-1}(x) = \frac{1}{x^2} \quad , \quad l > 0 \quad . \tag{18.23}$$

If one wants to calculate the regular spherical Bessel functions accurately for small values of the argument, one has to forsake the recursion formula (18.17) and work with other methods.

19. Scattering of an Uncharged Particle from a Spherically Symmetric Potential

19.1 Formulation of the Problem

With the methods considered in Chaps. 13, 17 and 18 we are now in a position to calculate a differential scattering cross-section. We shall consider a simple optical model for the elastic scattering of neutrons by medium or heavy atomic nuclei.

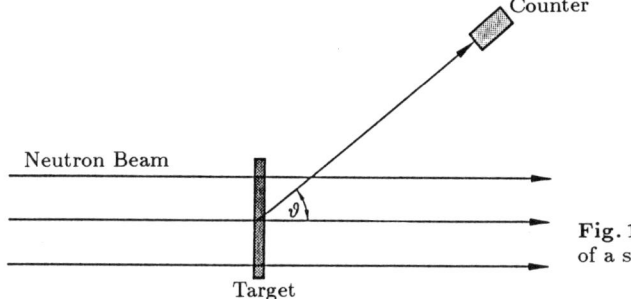

Fig. 19.1. Schematic arrangement of a scattering experiment

The physical situation is sketched in Fig. 19.1. With a particle accelerator and a first nuclear reaction one produces a monoenergetic neutron beam. In a target chamber the neutron beam falls on a thin foil, which contains the target nuclei. When a neutron hits a target nucleus it gets deflected from the original direction by the scattering angle ϑ. A neutron counter is mounted on a swivel arm to count the scattered neutrons. The number measured over a certain period of time depends on the scattering angle ϑ, on the intensity of the neutron beam, on the density of target nuclei on the foil and on the solid angle $d\Omega$ of the counter aperture relative to the scattering centre. If neither the neutrons nor the target nuclei are polarised, then the scattering process is symmetrical with regard to rotation about the axis of the beam, i.e. it is independent of the azimuthal angle φ.

What interests us is the probability that a neutron gets deflected by an angle ϑ. As a measure of this probability a conceptual quantity has been introduced, which is called the differential scattering cross-section. A small diaphragm with the aperture $d\sigma$ is placed in the incident neutron beam. When this diaphragm cuts out just as many neutrons as are scattered by a target nucleus into the counter with an aperture of solid angle $d\Omega$, then we have

$$\frac{d\sigma}{d\Omega} = \text{differential scattering cross-section} \,. \tag{19.1}$$

The differential scattering cross-section is a function of the scattering angle ϑ and depends also on the energy of the incident neutrons and on the nature of the target nuclei. With light nuclei it also depends on whether one chooses as reference system the laboratory system, in which the target nuclei are at rest, or the centre of mass system, in which the centre of mass of the neutron-nucleus system is at rest. Since we are considering medium or heavy target nuclei, this difference does not matter.

The scattering process can be described theoretically by the Schrödinger equation,

$$-\frac{\hbar^2}{2m}\Delta\psi(\mathbf{r}) + V(r)\psi(\mathbf{r}) = E\psi(\mathbf{r}) \quad . \tag{19.2}$$

The vector $\mathbf{r}$ denotes the position of the neutron relative to the target nucleus. The spherically symmetric potential $V(r)$ represents the effect of the nucleus on the neutron. We choose for $V(r)$ a Woods-Saxon potential,

$$V(r) = \frac{-V_0}{1 + \exp(\frac{r-R}{a})} \quad , \tag{19.3}$$

which has the form shown in Fig. 19.2.

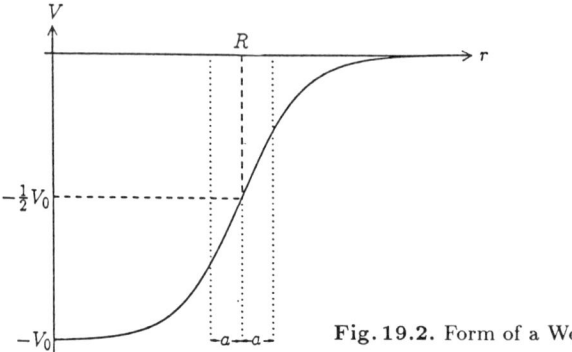

Fig. 19.2. Form of a Woods-Saxon potential

For a realistic analysis of a measured differential cross-section the potential should also have an imaginary part as well as a spin-orbital term. However, we limit ourselves to the potential (19.3) and shall thereby already obtain quite a reasonable differential scattering cross-section.

As already mentioned several times, one needs boundary conditions in order to determine the solution of the Schrödinger equation (19.2). Equivalent to the provision of boundary conditions, however, is the specification of the asymptotic form of the wave function $\psi(r)$. This alternative possibility enables us to introduce information on the experimental set-up into the mathematical description.

As sketched in Fig. 19.3, we describe the neutron beam by a plane wave $\exp(\mathrm{i}kz)$. This plane wave is an eigenstate of the momentum, where the momen-

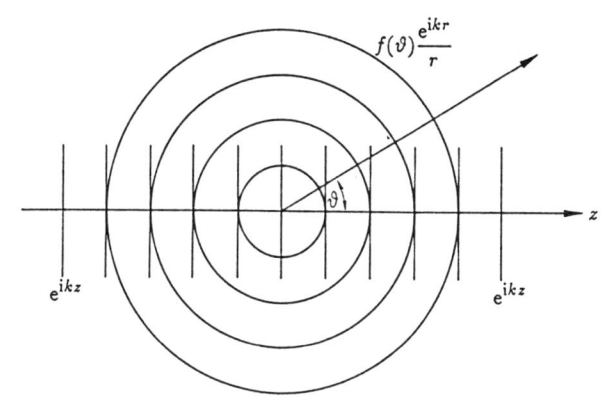

Fig. 19.3. Asymptotic form of a scattering solution

tum vector lies in the direction of the z-axis. The potential can modify the plane wave only by producing a scattering wave. The scattered wave must describe an outgoing neutron, i.e. with respect to r it must have the form $\exp(ikr)/r$. The amplitude f of the scattered wave may depend on the scattering angle ϑ. We therefore require that the wave function $\psi(r)$ have the asymptotic form

$$\psi(\mathbf{r}) \to e^{ikz} + f(\vartheta)\frac{e^{ikr}}{r} \quad \text{for} \quad r \to \infty \tag{19.4}$$

where k is the wave number introduced in (18.5). This ensures that each incoming neutron comes from the neutron source and that each outgoing neutron has either been scattered or has passed the target without being deflected. The angle φ does not appear in the scattering amplitude, because the spherically symmetric potential $V(r)$ can give the neutron no angular momentum about the z-axis.

From the asymptotic form (19.4) of the wave function $\psi(r)$ one can deduce the differential scattering cross-section. Since the absolute square of the wave function is proportional to the neutron probability density one obtains

$$\frac{d\sigma}{d\Omega} = |f(\vartheta)|^2 \;. \tag{19.5}$$

Here we make the usual assumption that in the asymptotic region the interference terms between the scattered wave and the plane wave can be ignored, since they arise from the idealisation of the neutron beam by a plane wave of infinite extent. We therefore obtain the differential scattering cross-section if we can manage to find a solution $\psi(r)$ of (19.2) which has the asymptotic form (19.4).

In the previous chapters we have learnt how to calculate partial solutions of the Schrödinger equation (19.2) in polar coordinates (r, ϑ, φ), with and without a potential $V(r)$. We now have to superpose the partial solutions to obtain a resultant solution with the asymptotic form (19.4).

19.2 Mathematical Treatment of the Scattering Problem

We start from the partial wave decomposition (18.2) of the wave function $\psi(r)$ and substitute the asymptotic form (18.13) for the radial functions $u_l(r)$. We thus obtain the expansion

$$\psi(\mathbf{r}) \to \sum_{l,m} c_{lm} \frac{\sin(kr - l\pi/2 + \delta_l)}{kr} Y_{l,m}(\vartheta, \varphi), \tag{19.6}$$

which is valid in the asymptotic region. The expansion coefficients c_{lm} are determined by boundary conditions. The boundary conditions valid for our scattering system are contained in (19.4), i.e. we have to choose the coefficients c_{lm} so that the wave functions (19.4) and (19.6) coincide asymptotically. We achieve this by expanding $\psi(\mathbf{r})$ of (19.4) into partial waves and equating coefficients.

First of all one sees that all the c_{lm} with $m \neq 0$ vanish, since $\psi(\mathbf{r})$ in (19.4) depends only on r and ϑ, but not on φ. We therefore only have to calculate the coefficients c_{l0}.

The expansion of the plane wave in (19.4) into spherical waves reads [19.1]

$$e^{ikz} = e^{ikr\cos\vartheta} = \sum_{l=0}^{\infty}(2l+1)i^l j_l(kr) P_l(\cos\vartheta) \ . \tag{19.7}$$

From (18.11) we have, for $r \to \infty$,

$$j_l(kr) \to \frac{\sin(kr - l\pi/2)}{kr} = \frac{1}{2ikr}(e^{i(kr-l\pi/2)} - e^{-i(kr-l\pi/2)})$$

$$= \frac{1}{2ikr}((-i)^l e^{ikr} - i^l e^{-ikr}) \ . \tag{19.8}$$

We can therefore write, for $r \to \infty$,

$$e^{ikz} \to \frac{1}{2ikr}\sum_{l=0}^{\infty}(2l+1)(e^{ikr} - (-1)^l e^{-ikr}) P_l(\cos\vartheta) \ . \tag{19.9}$$

The scattered wave in (19.4) can also be decomposed into partial waves, by expanding $f(\vartheta)$ in Legendre polynomials. We write

$$f(\vartheta) = \frac{1}{2ik}\sum_{l=0}^{\infty}(2l+1) b_l P_l(\cos\vartheta) \ , \tag{19.10}$$

where, in order to simplify the subsequent calculation, we have taken the factor $(2l+1)/2ik$ out of the coefficients b_l. Altogether we obtain, for $r \to \infty$,

$$\psi(\mathbf{r}) \to e^{ikz} + f(\vartheta)\frac{e^{ikr}}{r}$$

$$= \frac{1}{2ikr}\sum_{l=0}^{\infty}(2l+1)\{(1+b_l)e^{ikr} - (-1)^l e^{-ikr}\} P_l(\cos\vartheta) \ . \tag{19.11}$$

Because of $c_{lm} = 0$ for $m \neq 0$ and $Y_{l,0}(\vartheta,\varphi) = P_l(\cos\vartheta)$ we can write (19.6) in the following form

$$\psi(\mathbf{r}) \to \sum_l c_{l0} \frac{\sin(kr - l\pi/2 + \delta_l)}{kr} P_l(\cos\vartheta)$$

$$= \sum_l c_{l0} \frac{1}{2ikr}\left\{(-i)^l e^{ikr} e^{i\delta_l} - i^l e^{-ikr} e^{-i\delta_l}\right\} P_l(\cos\vartheta)$$

$$= \frac{1}{2ikr} \sum_l c_{l0}(-i)^l e^{-i\delta_l}\left\{e^{2i\delta_l} e^{ikr} - (-1)^l e^{-ikr}\right\} P_l(\cos\vartheta) \quad . \quad (19.12)$$

Comparing (19.11) with (19.12) and equating coefficients we obtain

$$c_{l0} = (2l+1)i^l e^{i\delta_l}, \quad b_l = e^{2i\delta_l} - 1 \quad . \tag{19.13}$$

From (19.10) it then follows that

$$f(\vartheta) = \frac{1}{2ik} \sum_{l=0}^{\infty}(2l+1)(e^{2i\delta_l} - 1)P_l(\cos\vartheta)$$

$$= \frac{1}{k} \sum_{l=0}^{\infty}(2l+1)e^{i\delta_l} \sin(\delta_l) P_l(\cos\vartheta)$$

$$= \frac{1}{k} \sum_{l=0}^{\infty}(2l+1)(\cos(\delta_l)\sin(\delta_l) + i \sin^2(\delta_l)) P_l(\cos\vartheta)$$

$$= \frac{1}{2k} \sum_{l=0}^{\infty}(2l+1) \sin(2\delta_l) P_l(\cos\vartheta)$$

$$+ \frac{i}{k} \sum_{l=0}^{\infty}(2l+1) \sin^2(\delta_l) P_l(\cos\vartheta) \quad .$$

$$\tag{19.14}$$

According to (19.5) we then obtain as differential scattering cross-section

$$\frac{d\sigma}{d\Omega} = \frac{1}{k^2}\left|\sum_{l=0}^{\infty}(2l+1)e^{i\delta_l} \sin(\delta_l) P_l(\cos\vartheta)\right|^2 , \tag{19.15}$$

or, written out in detail,

$$\frac{d\sigma}{d\Omega} = (\text{Re}\,[f(\vartheta)])^2 + (\text{Im}\,[f(\vartheta)])^2 \quad , \quad \text{with} \tag{19.16}$$

$$\text{Re}\,[f(\vartheta)] = \frac{1}{2k} \sum_{l=0}^{\infty}(2l+1) \sin(2\delta_l) P_l(\cos\vartheta) \quad \text{and} \tag{19.17a}$$

$$\text{Im}\,[f(\vartheta)] = \frac{1}{k} \sum_{l=0}^{\infty}(2l+1) \sin^2(\delta_l) P_l(\cos\vartheta) \quad . \tag{19.17b}$$

A further physically interesting quantity is the total cross-section σ, which is obtained by integrating the differential scattering cross section over the solid angle. The total cross-section can be understood as the area which the target nucleus cuts from the incident beam.

For the calculation of the total scattering cross-section σ we need the orthogonality relation of the Legendre polynomials (see Chap. 3 and [19.2]):

$$\int_{-1}^{1} P_l(\cos\vartheta) P_{l'}(\cos\vartheta) d(\cos\vartheta) = \frac{2}{2l+1}\delta_{ll'} \quad . \tag{19.18}$$

Then we obtain from (19.15):

$$\begin{aligned}
\sigma &= \int \frac{d\sigma}{d\Omega}(\vartheta) d\Omega \\
&= \int_0^{2\pi} d\varphi \int_{-1}^{1} \frac{d\sigma}{d\Omega}(\vartheta) d(\cos\vartheta) = 2\pi \int_{-1}^{1} \frac{d\sigma}{d\Omega}(\vartheta) d(\cos\vartheta) \\
&= \frac{2\pi}{k^2} \int_{-1}^{1} \left| \sum_{l=0}^{\infty} (2l+1) e^{i\delta_l} \sin(\delta_l) P_l(\cos\vartheta) \right|^2 d(\cos\vartheta) \\
&= \frac{2\pi}{k^2} \int_{-1}^{1} \sum_{l,l'=0}^{\infty} (2l+1)(2l'+1) e^{i(\delta_l - \delta_{l'})} \sin(\delta_l) \sin(\delta_{l'}) \\
&\quad \times P_l(\cos\vartheta) P_{l'}(\cos\vartheta) d(\cos\vartheta) \\
&= \frac{4\pi}{k^2} \sum_{l=0}^{\infty} (2l+1) \sin^2(\delta_l) \quad .
\end{aligned} \tag{19.19}$$

For the total cross-section the contributions of the individual partial waves are additive, whereas for the differential cross-section interference terms occur. We can interpret this in the following way: for each particular angle there is a contribution to the cross-section which arises from the interference of different partial waves. These interference terms cancel out on averaging over solid angles, since partial waves with different angular momentum quantum numbers are orthogonal to each other.

If we compare (19.19) and (19.17b) and take account of the fact that for all $l \geq 0$

$$P_l(1) = 1 \tag{19.20}$$

[19.2], then we recognise a relation known as the "Optical Theorem":

$$\sigma = \frac{4\pi}{k} \text{Im}\,[f(0)] \quad . \tag{19.21}$$

19.3 Programming

In our program for calculating the scattering cross-section we can use several subprograms already constructed in earlier chapters. We shall make use of the subprograms BESJ, BESN (both from Chap. 18) and PLM (from Chap. 17), together with the FUNCTION subprogram FAK (from Chap. 16).

In order to calculate the potential from (19.3) we write the FUNCTION subprogram V (Fig. 19.5). We use the following potential parameters [19.3]:

$$V_0 = V_1 - V_2 E - \left(1 - \frac{2Z}{A}\right) V_3 \quad , \tag{19.22a}$$

$$V_1 = 56.3 \,\text{MeV} \quad , \quad V_2 = 0.32 \quad , \quad V_3 = 24.0 \,\text{MeV} \quad , \tag{19.22b}$$

$$R = r_0 A^{1/3} \quad \text{with} \quad r_0 = 1.17 \,\text{fm} \quad , \tag{19.22c}$$

$$a = 0.75 \,\text{fm} \quad . \tag{19.22d}$$

Here A is the mass number and Z the charge number of the target nucleus, and E is the kinetic energy of the incident neutron. The parameter values are valid in the region $E \leq 50 \,\text{MeV}$ and $A \geq 40$.

The FUNCTION subprogram V is seen in Figs. 19.4 and 19.5. Parameter values which are kept fixed are specified in a PARAMETER instruction (line 103). The IF-block (lines 107 to 111) prevents a possible exponent overflow.

Physical title	FORTRAN title	Physical title	FORTRAN title
V_1	V1	E	E
V_2	V2	A	A
V_3	V3	Z	Z
a	A0	r	R
r_0	R0	R	R1
V_0	V0		

Fig. 19.4. Notation employed in FUNCTION subprogram V

```
DOUBLE PRECISION FUNCTION V(R,E,A,Z)                            100

IMPLICIT DOUBLE PRECISION (A-H,O-Z)                             101
INTEGER A, Z                                                    102
PARAMETER (V1=56.3D0, V2=0.32D0, V3=24.D0, R0=1.17D0, A0=0.75D0) 103
PARAMETER (EXPMAX=700.D0)                                       104
V0=V1-V2*E-V3*(1.D0-(2.D0*Z)/A)                                 105
R1=R0*A**(1.D0/3.D0)                                            106
IF((R-R1)/A0.LE.EXPMAX) THEN                                    107
  V=-V0/(1.D0+EXP((R-R1)/A0))                                   108
ELSE                                                            109
  V=0.D0                                                        110
ENDIF                                                           111
END                                                             112
```

Fig. 19.5. FUNCTION subprogram V

Physical title	FORTRAN title	Physical title	FORTRAN title
h_b	HB	r_i	RR(I)
b	B	$V(r_i)$	VV(RR(I))
i_b	IB	$w(r_{i+1})$	W2
h_π	HPI	$u(r_{i+1})$	U2
π	PI	$w(r_i), u(r_i)$	W1, U1
i_π	IPI	$w(r_{i-1}), u(r_{i-1})$	W0, U0
$\hbar$	HBAR	$j_l(x)$	BESJ(L,X)
$\hbar^2/2m$	H2M	$n_l(x)$	BESN(L,X)
m_N	MN	δ_l	DELTA(L)
A	A	$P_l(\cos x)$	PLM(X,L,0)
Z	Z	ϑ_i	THETA(I)
k	K	$P_l(\cos \vartheta_i)$	PLTH
E	E	$\mathrm{Re}[f(\vartheta_i)]$	REF
l	L	$\mathrm{Im}[f(\vartheta_i)]$	IMF
l_P	LP	$d\sigma/d\Omega(\vartheta_i)$	DSIGMA(I)
$l_{\max}$	LMAX	σ	SIGMA

Fig. 19.6. Notation used in main program KAP19

Figure 19.7 shows the numerical portion of the main program; the notation employed is displayed in Fig. 19.6. As mentioned at the outset, we are considering the scattering of a neutron by a medium or heavy atomic nucleus. In this case the reduced mass m is not very different from the neutron mass m_N. We shall nevertheless calculate with the reduced mass

$$m = \frac{m_N m_K}{m_N + m_K} \quad . \tag{19.23}$$

For the mass m_K of the nucleus we shall use the approximation $m_K = A[u]$ (u is the atomic unit of mass). The neutron mass m_N in atomic units and $\hbar$ in the units (MeVu)$^{1/2}$ are specified in a PARAMETER instruction. We accordingly obtain the constant $\hbar^2/(2m)$ in line 208 and the wave number from (18.5) in line 209. For the solution of the radial Schrödinger equation we use the method developed in Chap. 13. In the interval $[0, b]$ we introduce i_b mesh points with mesh width h_b (line 212). The integration of the radial Schrödinger equation by the Fox-Goodwin method follows in lines 220 to 227, see (13.3) and (13.31). Since we need the solution $u_l(r)$ only at the last two mesh points b and $(b-h_b)$, we do not use the subprogram FOX, which would produce the function values at all the mesh points. In order to economise in computation time we calculate the potential values at the mesh points only once and store them in the array VV(I), see line 213.

After the execution of the Fox-Goodwin recursion the function values $u_l(b-h_b)$ and $u_l(b)$ are available, denoted by U0 and U1. From these values the scattering phase shift δ_l is calculated according to (18.10), using the FUNCTION

```
      PROGRAM KAP19                                                200

      IMPLICIT DOUBLE PRECISION (A-H,O-Z)                          201
      DOUBLE PRECISION IMF, K, MN                                  202
      INTEGER A, Z                                                 203
      PARAMETER (LMAX=18, IPI=90, PI=3.1415926535D0, IB=500, B=20.D0)  204
      PARAMETER (HBAR=6.4655D0, MN=1.008665D0)                     205
      DIMENSION VV(IB),RR(IB)                                      206
      DIMENSION THETA(0:IPI), DSIGMA(0:IPI), DELTA(0:LMAX)         207

*     (Input: A, Z, E, LP)

      H2M=HBAR*HBAR*(MN+A)/(2.D0*MN*A)                             208
      K=SQRT(E/H2M)                                                209
      HB=B/IB                                                      210
      DO 10 I=1,IB                                                 211
       RR(I)=I*HB                                                  212
       VV(I)=V(RR(I),E,A,Z)                                        213
10    CONTINUE                                                     214
      DO 30 L=0,LP                                                 215
       W0=0.D0                                                     216
       W1=(E-VV(1))/H2M-L*(L+1)/(HB*HB)                            217
       U0=0.D0                                                     218
       U1=1.D0                                                     219
       DO 20 I=1,IB-1                                              220
        W2=(E-VV(I+1))/H2M-L*(L+1)/(RR(I+1)*RR(I+1))               221
        U2=(2.D0*U1-U0-HB*HB/12.D0*(10.D0*W1*U1+W0*U0))            222
     &    /(1.D0+HB*HB/12.D0*W2)                                   223
        W0=W1                                                      224
        W1=W2                                                      225
        U0=U1                                                      226
        U1=U2                                                      227
20     CONTINUE                                                    228
       A1=(B-HB)*BESJ(K*(B-HB),L)*U1-B*BESJ(K*B,L)*U0              229
       A2=(B-HB)*BESN(K*(B-HB),L)*U1-B*BESN(K*B,L)*U0              230
       DELTA(L)=ATAN(A1/A2)                                        231
30    CONTINUE                                                     232
      HPI=PI/IPI                                                   233
      DO 50 I=0,IPI                                                234
       THETA(I)=I*HPI                                              235
       REF=0.D0                                                    236
       IMF=0.D0                                                    237
       DO 40 L=0,LP                                                238
        PLTH=PLM(THETA(I),L,0)                                     239
        REF=REF+0.5D0/K*(2.D0*L+1.D0)*SIN(2.D0*DELTA(L))*PLTH      240
        IMF=IMF+ 1.D0/K*(2.D0*L+1.D0)*(SIN(DELTA(L)))**2*PLTH      241
40     CONTINUE                                                    242
       DSIGMA(I)=(REF*REF+IMF*IMF)                                 243
       IF(I.EQ.0)  SIGMA=4.D0*PI/K*IMF                             244
50    CONTINUE                                                     245

*     (Output: DSIGMA, SIGMA)

      END                                                          246
```

Fig. 19.7. Numerical portion of main program KAP19

subprograms BESJ and BESN (lines 229 to 231). The calculation of the radial wave function as well as the calculation of the scattering phase shift are carried out within a DO-loop (lines 215 to 232) for all angular momentum quantum numbers from $l = 0$ to $l = l_P$.

The calculation of the real and imaginary parts of the scattering amplitude $f(\vartheta)$ from (19.17) follows in lines 240 and 241. The scattering angle ϑ is discretised by i_π mesh points in the interval $[0, \pi]$. The mesh width $h_\pi = \pi/i_\pi$ is calculated in line 233. The Legendre polynomials occurring in (19.17) are calculated by calling the FUNCTION subprogram PLM (line 239). The program PLM is suitable for calculating not only the associated Legendre functions $P_l^m(\cos\vartheta)$, but also the Legendre polynomials $P_l(\cos\vartheta) \equiv P_l^0(\cos\vartheta)$. The infinite sum in (19.17) is truncated at $l = l_P$, where l_P may not be greater than the maximal angular momentum quantum number $l_{\max}$ used in dimensioning the arrays.

From the scattering amplitude, using (19.16) and (19.21), one finally obtains the differential scattering cross-section $d\sigma/d\Omega$ and the total cross-section σ (lines 243 and 244).

By the PARAMETER instruction in lines 204 and 205 the following quantities are specified: the upper interval boundary $b = 20\,\text{fm}$, the number of sub-intervals $i_b = 500$ in the interval $[0, b]$, the number of sub-intervals $i_\pi = 90$ in the interval $[0, \pi]$, the maximal angular momentum quantum number $l_{\max} = 18$, the constant $\hbar = 6.4655\,\text{fm}\,(\text{MeVu})^{1/2}$ and the neutron mass $m_N = 1.008665u$. Supplied as input are the mass number A and the nuclear charge number Z of the target nucleus, the energy E of the neutron and the angular momentum quantum number l_P, at which the summation over angular momentum partial waves is truncated. As output one obtains the differential cross-section $d\sigma/d\Omega$ and the total cross-section σ. The output of $d\sigma/d\Omega$ as a function of the scattering angle ϑ is obtained in graphical form and optionally also in numerical form. Two diagrams are shown in the graphical output: in the first $d\sigma/d\Omega$ is plotted linearly, in the second logarithmically. All angles are converted from radians to degrees, and the cross-sections are converted from fm^2 to the familiar millibarn units (mb). A millibarn corresponds to $0.1\,\text{fm}^2$.

The program here presented in summary is to be found on the diskette in the file KAP-19.FOR.

19.4 Exercises

19.4.1 Take various atomic nuclei, and for various energies E find the least values of l_P, for which the results become stable.

19.4.2 Calculate the differential scattering cross-sections for various medium and heavy nuclei and various neutron energies. What trends in relation to the energy E and the mass number A can you identify?

19.4.3 Assuming the calculated cross-sections to be measured quantities, look for resonances. Can you determine the angular momentum of a resonant state?

19.5 Solutions to the Exercises

19.5.1 One finds that, with increasing values of l_P, the differential cross-section does no longer change its shape when the total cross-section does no longer change its value. In testing the convergence one therefore only has to consider the total cross-section. With a medium heavy target nucleus and energies below 10 MeV it suffices to have $l_P = 5$. For heavy nuclei and an energy of 50 MeV one needs $l_P = 16$.

19.5.2 For low energies the differential cross-section has only a few maxima and minima. This is due to the fact that only a few partial waves contribute to the scattering cross-section in this case. With increasing energy the number of maxima and minima at first increases. Then for high energies the differential cross-sections again assume a simple shape. They show a pronounced maximum in the forward direction and a rapid decrease at larger scattering angles. This trend is the more pronounced, the heavier the target nucleus is. Figure 19.8 shows the differential cross-section for the scattering of neutrons by $^{56}_{26}$Fe at an incident energy of 10 MeV. Figure 19.9 shows the differential cross-section for the scattering of neutrons by $^{238}_{92}$U with a bombarding energy of 50 MeV; the rapid decrease of the differential cross-section with increasing scattering angle can be clearly recognised.

19.5.3 One finds resonances by plotting the excitation function, i.e. the total cross-section as a function of the energy. The calculated total cross-section for the scattering of neutrons by $^{56}_{26}$Fe, for example, has a sharp maximum at 2.776 MeV. There is a resonance here with a width of about 0.05 MeV. The associated differential cross-section is shown in Fig. 19.10.

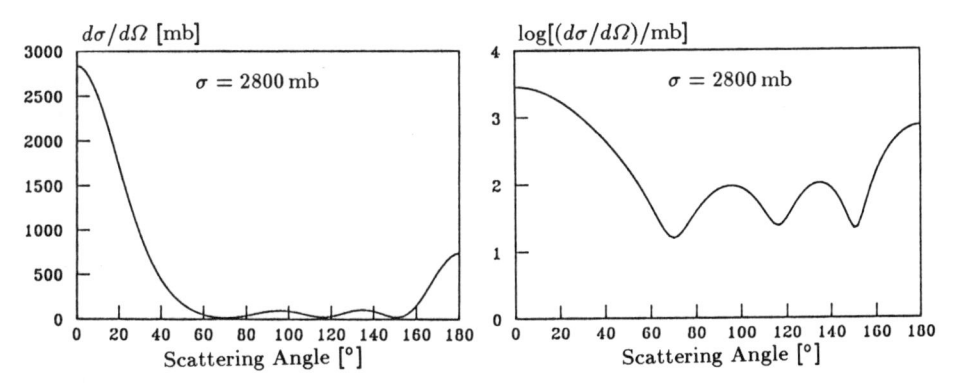

Fig. 19.8. Differential and total cross-section for the scattering of 10 MeV neutrons from $^{56}_{26}$Fe ($l_P = 5$)

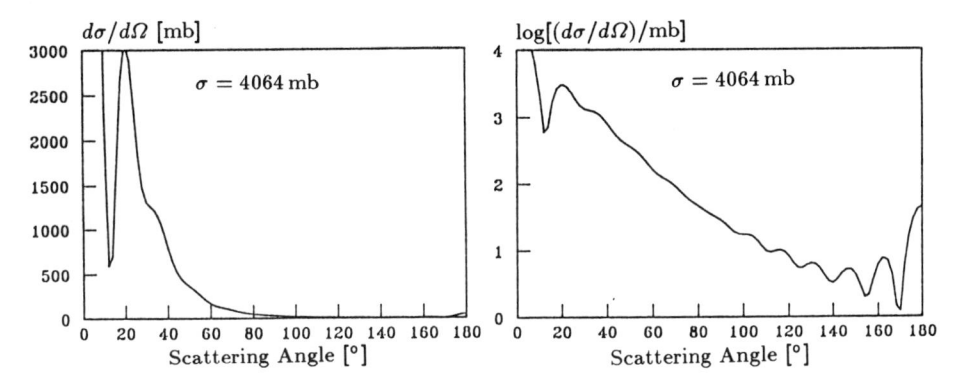

Fig. 19.9. Differential and total cross-section for the scattering of 50 MeV neutrons from $^{238}_{92}$U ($l_P = 16$)

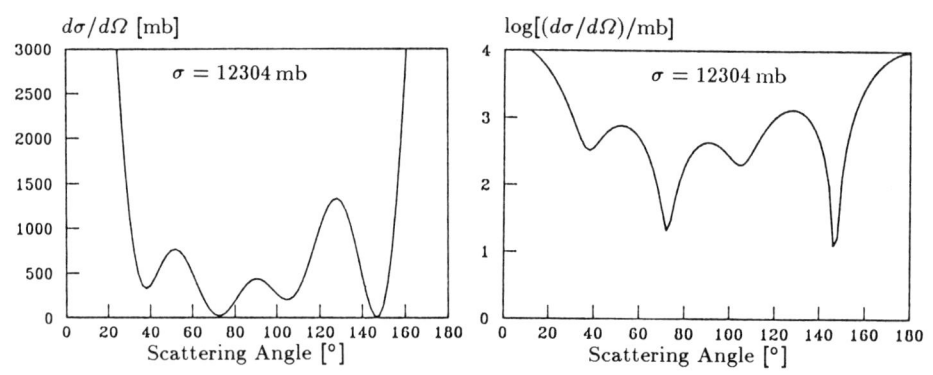

Fig. 19.10. Differential cross-section for the scattering of neutrons from $^{56}_{26}$Fe at the resonance energy $E = 2.776$ MeV

Whereas the differential cross-sections have only two minima for energies in the wider region about 2.776 MeV, in the immediate neighbourhood of 2.776 MeV there are 4 minima. In the region of the resonance, therefore, a Legendre polynomial with 4 zeros must make an essential contribution to the scattering. From this one concludes that the resonant state has the orbital angular momentum $l = 4$.

References

Chapter 3

3.1 E. Isaacson, H.B. Keller: *Analysis of Numerical Functions* (John Wiley and Sons, Inc., New York 1966)
3.2 A. Erdelyi, W. Magnus. F. Oberhettinger, F. Tricomi: *Higher Transcendental Functions*, Vols. 1 and 2 (McGraw-Hill, New York 1953)
3.3 I.S. Gradshteyn, I.M. Ryzhik: *Tables of Integrals, Series and Products*, corrected and enlarged edition (Academic Press, New York 1980)
3.4 M. Abramowitz, I.A. Stegun (eds.): *Handbook of Mathematical Functions*, 7th ed. (Dover Publications, New York 1970)

Chapter 5

5.1 E. Isaacson, H.B. Keller: *Analysis of Numerical Functions* (John Wiley and Sons, Inc., New York 1966)

Chapter 8

8.1 S.F. Dermott: Nucl. Phys. A**416**, 489c (1984)

Chapter 9

9.1 G. Forsythe, W. Wasow: *Finite Difference Methods for Partial Differential Equations*, 3rd ed. (Wiley, New York 1965)

Chapter 10

10.1 G. Kortüm: *Einführung in die chemische Thermodynamik* (Introduction into chemical thermodynamics), 6th ed. (Vandenhoeck and Ruprecht, Göttingen 1972)
10.2 D'Ans; Lax: *Taschenbuch für Chemiker and Physiker* (Handbook for chemists and physicists), Vol. 1, 2nd ed. (Springer, Berlin, Heidelberg 1967)

Chapter 12

12.1 A. Sommerfeld: *Theoretische Physik* (Theoretical Physics), Vol. 2, Reprint of 8th ed. (Harri Deutsch, Thun 1977)

Chapter 13

13.1 Apart from a trivial change for V_0 we take the values from B. Buck, H. Friedrich, C. Wheatley: Nucl. Phys. A**275**, 246 (1977)

13.2 B. Alder, S. Fernbach, M. Rothenberg (eds.), *Methods in Computational Physics*, Vol. 6, Nuclear Physics (Academic Press, New York 1966)
13.3 M. Abramowitz, I.A. Stegun (eds.): *Handbook of Mathematical Functions*, 7th ed. (Dover Publications, New York 1970)

Chapter 14

14.1 G. Eder: *Nuclear Forces* (M.I.T. Press, Cambridge, Massachusetts 1968)

Chapter 15

15.1 E. Schrödinger: Ann. Phys. **79**, 361 (1926)
15.2 W. Heisenberg: Z. Phys. **33**, 879 (1925)
15.3 G. Eder: *Nuclear Forces* (M.I.T. Press, Cambridge, Massachusetts 1968)
15.4 H.R. Schwarz, H. Rutishauser, E. Stiefel: *Numerical Analysis of Symmetric Matrices* (Prentice Hall, Englewood Cliffs, New Jersey 1974)
15.5 H. Rutishauser: Num. Math. **9**, 1 (1966)

Chapter 16

16.1 E.A. Hylleraas: Z. Phys. **54**, 347 (1929)
16.2 E.A. Hylleraas, B. Undheim: Z. Phys. **65**, 759 (1930)
16.3 H.R. Schwarz, H. Rutishauser, E. Stiefel: *Numerical Analysis of Symmetric Matrices* (Prentice Hall, Englewood Cliffs, New Jersey 1974)

Chapter 17

17.1 A.M. Messiah: *Quantum Mechanics I and II* (North-Holland, Amsterdam 1962)
17.2 I.S. Gradshteyn, I.M. Ryzhik: *Tables of Integrals, Series and Products*, corrected and enlarged edition (Academic Press, New York 1980)

Chapter 18

18.1 L.I. Schiff: *Quantum Mechanics*, 3rd ed. (McGraw-Hill, New York 1968)
18.2 A.M. Messiah: *Quantum Mechanics I and II* (North-Holland, Amsterdam 1962)

Chapter 19

19.1 L.I. Schiff: *Quantum Mechanics*, 3rd ed. (McGraw-Hill, New York 1968)
19.2 I.S. Gradshteyn, I.M. Ryzhik: *Tables of Integrals, Series and Products*, corrected and enlarged edition (Academic Press, New York 1980)
19.3 F.D. Becchetti Jr., G.W. Greenlees: Phys. Rev. **182**, 1190 (1969)

Subject Index

α-α scattering   139ff., 145, 148
Angular momentum barrier   see Centrifugal potential
Angular momentum operator   183

Bessel functions
— ordinary   192
— spherical   190–193, 195ff.
Bisection method   106
Boiling temperatures   108–114
Bound states, quantum mechanical   137–139, 145ff., 163ff.
Boundary value problem   90–92, 117ff., 137–140, 165ff.

Celestial mechanics   59–63, 81–85
Centrifugal potential   137, 146, 152, 156
Code page   6
Commentary   16ff., 19
Condensation   see Vapour pressure
Coulomb interaction   145, 147
Cowell method   see Fox-Goodwin method
Critical point of gases   104
CURSOR (subroutine)   6, 19
Cylindrical coordinates   91

Datafile   31
— internal   32
Declaration   see Type declaration
Derivative, numerical   see Differentiation, numerical
Differential equations, ordinary
— methods of solution   37ff., 44–46, 63–66, 140ff.
— reduction of order   36ff.
Differential equations, partial
— methods of solution   90ff., 117–119
— separation of variables   181–183
Differentiation, numerical   13ff., 19ff.
Dispersion, of wave velocity   126ff.

Earth-moon system   61, 76
Earthquake-proof building   54ff.

Eigenvalue problem
— for symmetrical matrices   159ff.
— generalized   168, 172–175
Electromagnetic waves   see Light waves
Electron lens   90ff., 101
Electrostatic field   90ff.
Equipotential lines   97–100
Euler method
— improved   44–45
— ordinary   37ff., 41

Flight path of a spacecraft   59–63, 79ff.
Floating point numbers   12
— exponent overflow   120
— precision   12
— range of exponent   12
FORTRAN norm   3, 5
Fourier series   115, 118ff., 122ff.
Fox-Goodwin method   141ff., 144ff.
Friction
— sliding   35
— static   35
Frictional forces   35ff.

Gas equation
— ideal   102, 112ff.
— realistic (van der Waals)   102–106, 112ff.
Gauss-Legendre integration   25–29, 33ff.
Gaussian error function   145, 148
GBOX (graphics subroutine)   100
GCHART (graphics subroutine)   133ff.
GCIRCL (graphics subroutine)   77ff.
GCLOSE (graphics subroutine)   40
GCOLOR (graphics subroutine)   41
GDRAW (graphics subroutine)   41, 57
Geo-power station   115–117, 122ff.
GINIT (graphics subroutine)   77ff.
GINKEY (graphics subroutine)   77
GINPUT (graphics subroutine)   112
GLINE (graphics subroutine)   40
GMARK (graphics subroutine)   77
GMOVE (graphics subroutine)   40, 57
GOPEN (graphics subroutine)   39
Graphics Library
— Choosing the graphics mode   8ff.
— Installation   2ff., 6, 10ff.

— Necessary hardware 1ff., 6–8
— Necessary software 1ff., 5–10
— Setting the palette of colours 9ff.
— Versions 2ff., 5–11
Graphics subroutines
— Close graphics 40, 76ff.
— Colour selection 41
— Draw boxes 100
— Draw circles 77ff.
— Draw curves 40, 57
— Draw frames and axes 133ff.
— Draw lines 40ff.
— Draw marks 77
— Initialize graphics 39ff., 76ff.
— Input by cross wires 112
— Input of characters 77
— Input of numbers 133
— Output of text 78, 111
— Redefinition of windows 133
Group velocity 126ff., 134ff.
GVALUE (graphics subroutine) 133ff.
GWINDO (graphics subroutine) 133
GWRITE (graphics subroutine) 78, 111

Hamilton equations 59–63, 82–84
Hamilton function 60, 62ff., 82
Hamilton operator 136, 158ff., 165, 171ff.
Hardware prerequisites 1ff., 6–8
Heat conduction equation 117
Helium atom 165–172, 180
Horseshoe orbit 84, 89
Hot Dry Rock Process 115–117
Hylleraas method 165–172, 180

Input
— alphanumerical 15–19
— graphical see Graphics subroutines
Integration, numerical 22–29, 33ff.
Intervals, nested 106–107

Lagrange function 60, 62ff., 82
Laguerre polynomial, generalized 151
Laplace equation 90–93
Legendre polynomials
— associated 185ff.
— ordinary 27–29, 200–202
Libration oscillations 84, 89
Light waves 126ff.

Maxwell's line 103–106

Neutron scattering 197ff., 202, 206ff.
Newton-Cotes method 25
Noumerov method see Fox-Goodwin method

Optical Theorem 202
Oscillation
— anharmonic 43ff., 50–52
— coupled 53ff., 57ff.
— damped 35ff., 41ff., 54
— forced 43ff., 52, 54
— harmonic 35ff., 53ff., 57ff.
Oscillator
— classical see Oscillation
— quantum mechanical 149ff., 156ff.
Oscillator functions 150–153, 156, 158ff.
Output
— alphanumerical 15–19
— graphical see Graphics subroutines
Over-relaxation, successive 93ff., 100ff.

Pauli principle 139ff., 165
Pendulum see Oscillation
Perturbation theory, first order 180
Phase shift see Scattering phase shift
Phase velocity 125–128, 134ff.
Polar coordinates
— plane 61ff.
— spherical 136ff., 181–183
Probability density
— classical 152, 156ff., 188
— quantum mechanical 137–139, 156ff., 188
Programming methods 4ff., 16–18

Recursion formulae
— for associated Legendre polynomials 184ff.
— for differential equations see Differential equations
— for generalized Laguerre polynomials 151
— for harmonic oscillator functions 151, 153
— for spherical Bessel functions 191ff.
— for trigonometric functions 130
Resonances
— classical 52, 54ff., 85
— quantum mechanical 140, 148, 206ff.
Ritz variational principle 168, 180
Rodrigues formulae 26, 184
Rounding error 12, 20ff.
Runge-Kutta method 46ff., 49ff., 63–66, 79

Scattering amplitude 199, 200–202
Scattering cross-section
— differential 197–202, 206ff.
— total 202ff., 206ff.
Scattering experiment 197
Scattering phase shift 190ff., 200ff.
Scattering solutions 138, 159, 163ff.

Schrödinger equation
— angular part   181–183
— as matrix equation   158ff., 166, 168
— as operator equation   158
— boundary conditions   137–140, 165ff.
— for free particles   189ff.
— for the Helium atom   165ff.
— in configuration space   136, 181–183, 189
— radial part   136–140, 181ff., 189ff.
— separation   181–183
Simpson rule   24ff., 33ff., 160
Software prerequisites   1ff., 5–10
Spherical coordinates   136ff., 181–183
Spherical harmonics   181ff., 184
State Basis   158ff., 166–169
Superposition principle   125

Taylor series   13ff., 44
Three-body problem
— in celestial mechanics   60–63, 81–85
— quantum mechanical   see Helium atom
Three-point formula   14
Trapezoidal rule   23ff., 33ff., 130ff.
Trojan satellites, orbits of   84, 89
Two-point formula   14
Type declaration   29

Van der Waals equation   see Gas equation
Vapour pressure   103–106, 112–114

Water waves   127–129, 134ff.
Wave packet   125–129, 134ff.
Wave packets, quantum mechanical   127ff.
Wave propagation   125–129

S. Brandt, H. D. Dahmen, University of Siegen

Quantum Mechanics on the Personal Computer

1989. X, 267 pp. With a Program Diskette, 69 figs. and 284 exercises. Hardcover DM 98,– ISBN 3-540-51541-0

Quantum Mechanics on the PC gives the most up-to-date access to elementary quantum mechanics. Based on the interactive program Interquanta (included on a 5 1/4" Diskette, MS-DOS) and its extensive 3D color graphic features, the book guides the reader through computer experiments on

- free particles
- bound states and scattering from various potentials
- two particle problems
- properties of special functions of mathematical physics.

The course, with a wide variety of more than 200 detailed, class-tested problems, provides students with invaluable practical experience of complex probability amplitudes, eigenvalues, scattering cross sections and the like. Lecturers and teachers can use it to make excellent hands-on classroom demonstrations for their own courses.

Springer-Verlag Berlin Heidelberg New York London Paris Tokyo Hong Kong

Springer